Molecular Diagnostics

NIPA® GENX ELECTRONIC RESOURCES & SOLUTIONS P. LTD.
New Delhi-110 034

Molecular Diagnostics
A Practical Manual

Diwakar Singh, Ph.D., FSAB
K.P. Suthar, Ph.D.
Reetu Mehta, Ph.D., FSAB

NIPA® GENX ELECTRONIC RESOURCES & SOLUTIONS P. LTD.
New Delhi-110 034

NIPA® GENX ELECTRONIC
RESOURCES & SOLUTIONS P. LTD.

101,103, Vikas Surya Plaza, CU Block
L.S.C. Market, Pitam Pura, New Delhi-110 034
Ph : +91 11 27341616, 27341717, 27341718
E-mail: newindiapublishingagency@gmail.com
web: www.nipabooks.com

For customer assistance, please contact
Phone: + 91-11-27 34 17 17
Fax: + 91-11- 27 34 16 16
E-Mail: feedbacks@nipabooks.com

ISBN: 978-81-19235-22-3

Composed and Designed by NIPA®.

Preface

This practical manual on molecular diagnostics is written to help the students and faculties who are doing practical of molecular biology, biochemistry and biotechnology. This practical manual covers the syllabus of ICAR as per 5th Dean Committee for B.Sc. Agriculture students. This practical manual can also be used by the students of under graduate and post graduate of other streams of biological sciences. This manual explains step wise laboratory practical procedure for different experiments. This manual will also be helpful for those students and researchers who want to plan the experiment in molecular biology laboratory. In the manual several actual figures of results have been given for example. From these figures the students, researchers and faculty members can compare their observed results. Total seventeen different experiments have been explained including screening of abiotic stress tolerant & sensitive genotypes, isolation & quantification of protein, isolation & assay of Reactive Oxygen Species (ROS) enzymes, separation of isozymes & protein by Polyacrylamide Gel Electrophoresis (PAGE), DNA & RNA isolation, quantification and quality test by spectrophotometer and agarsoe gel electrophoresis, PCR, RAPD, southern blotting and ELISA technique. The authors are also engaged in research and teaching in this area and having wide experience in molecular biology. In the end of each experiment some questions have been asked to solve by the students that will help in better understanding of the topic. This manual also contains set of objective questions related to the molecular diagnostics that will be very useful for their competitive exams and regular theoretical exams as well. The molecular biology techniques are gaining much importance in life sciences and without this now a day's life science experiments cannot be thought. The suggestions and critics for further up gradation of this manual is always welcome.

We would like to thank to the publisher to bring this book in its final shape for publication.

Diwakar Singh, Ph.D., FSAB
K.P. Suthar, Ph.D.
Reetu Mehta, Ph.D., FSAB

BONAFIDE CERTIFICATE

Certified that this is the *bonafide* record of work done by the student Mr./Miss. ____________________________

of ____

college/department____________________________

Semester ______________ degree during the academic year ________________the course no. ______________

course title ____________________________________

Course Teacher **Head of Department**

Submitted for the university/college/department practical examination held on _____________ at ______________

Internal Examiner Date **External Examiner**

Date:

General Laboratory Rules

1. No eating, drinking, applying cosmetics, *etc.* in the lab.
2. Access to the laboratory is restricted only when the faculty is present and experiments are in progress.
3. Work area should be cleaned for any chemical spill.
4. Use only mechanical pipetting devices. Mouth pipetting is not permitted.
5. Hands should be washed before and after working in the laboratory.
6. Wear disposable gloves while working with hazardous chemicals such as ethidium bromide, SDS, etc. Lab coats must be worn in the lab to protect your clothes from contamination. Lab coats should not be worn outside of the laboratory.
7. Waste must be disposed of properly.
8. After completion of experiments the equipments should be cleaned and lab-wares should be arranged properly.
9. Properly label the stock solutions and lab-wares.
10. Always keep your laboratory manual and record all observations in the given format.
11. Keep your belongings at right place.

Contents

1

Diagnosis of Drought Tolerance in Crop Plants by PEG Method

Objective: To identify drought tolerant genotypes by PEG method in laboratory.

Background information:

Among different abiotic stresses, drought is considered to be most detrimental for agricultural production. A drought is a period of below-average precipitation in a given climatic region that results in prolonged shortages in water availability, whether atmospheric, surface water or ground water. A drought can last for several months or years. Drought has substantial impact on ecosystem and agricultural productivity.

Some plant species are better able to tolerate drought stress than others. Genetic variability for drought tolerance also exists within many plant species (Blum, *et al.* 1980). The efficient exploitation of that variability for crop improvement would be facilitated by a fast and accurate screening method for identifying drought tolerant genotypes (Clark and Mc Craig, 1982). Several methods utilizing seedlings to screen for drought tolerance have been developed using an osmoticum such as polyethylene glycol (PEG) (Singh, *et al.* 1984). The other synonyms of PEG are Poly ethylene oxide (PEO) and Poly oxyethylene (POE); these other chemicals depend on the molecular weight. Chemical structure of PEG may be written as $H"(O"CH_2"CH_2)_n"OH$ (Kahovec, *et al.* 2002). PEG is used in certain abiotic stress experiments to produce high degree of osmotic pressure, normally in the order of 'tens of atmosphere' PEG produces osmotic pressure due to its flexible nature and water soluble polymer. PEG also makes certain specific interactions with bio-molecules. Due to above properties PEG is used as a specific chemical to produce osmotic pressure in several biochemical and bio-membrane related experiments. For some genotypes, the seedling response apparently is a good indicator of the plant's reaction to drought stress later in its life cycle. However, early screening of seed plants via seedlings tests must still be considered as a method for identifying genotypes that will tolerate drought stress in the vegetative stage (Andersen, *et al.* 1987).

Materials

1. Four known drought tolerant and susceptible wheat genotypes
2. Deionized water
3. Germination chamber
4. Plastic tub
5. Styrofoam blocks
6. Germination blotter paper
7. Plastic wraps
8. Hoagland solution
9. PEG – 8000 (Polyethylene glycol)
10. Plexiglass holder
11. Tissue paper, etc.

Procedure

1. Four wheat genotypes with known response to drought tolerance were taken
   a. Genotype A – Spring wheat with low drought tolerance (DS)
   b. Genotype B – Spring wheat with high drought tolerance (DT)
   c. Genotype C – Spring wheat with low drought tolerance (DS)
   d. Genotype D – Durum wheat with high drought tolerance (DT)
2. Soak the seeds for two hours in deionized water on first day.
3. Place the seeds under dark condition in a germination chamber with a plexiglass framework supporting the seeds (approximately 0.5 cm.) over deionized water.
4. Place the seeds on germination blotter paper extending down into the water to maintain a constant exposure to moisture.
5. Maintain high chamber humidity by misting the inside and covering the container with plastic wrap.
6. On third day, when roots become 2-4 cm long and coleoptiles 0.5-1.5 cm long respectively, 10-14 seeds must be transferred to each plexiglass holder.

7. Maintain following plant growth condition:

   On third day the plant growth tubs should be placed in the environment to be maintained for the remaining growth period. The standard environment for the experiments should be with growth chamber having 15/20°C night/day temperatures having daylength of 16 hours. Humidity must fluctuate from 50 to 100% and light intensity must be 400 $\mu Ecm^{-2}sec^{-1}$.

8. Keep the plexiglass holders in plexiglass frame floated in nutrient solution (1/4 Hoaglands) at a constant level in a plastic tub using styrofoam blocks.
9. On seventh day perform thinning to maintain 10 uniform plants and take the first weight followed by changing the nutrient solution.
10. In order to take the weight remove the holder with plants from the frame work and blot the holder and roots with tissue paper, take the weight and return to the tub with nutrient solution.
11. On tenth day take second weight and replace the nutrient solution with 20% PEG-8000.
12. On fourteenth day take third weight and replace the nutrient solution with 20% PEG-8000.
13. Take final weight on 18th day of growth.

Observation

Genotype	Fresh wt. in gm. at 7 DAS (% Growth of Day 10)	Fresh wt. in gm. at 10 DAS (% Growth of Day 10)	Fresh wt. in gm. at 14 DAS (% Growth of Day 10)	Fresh wt. in gm. at 18 DAS (% Growth of Day 10)
Genotype A (DS)	——	——	——	——
Genotype B (DT)	——	——	——	——
Genotype C (DS)	——	——	——	——
Genotype D (DT)	——	——	——	——

Calculation

% Growth of Day 10 = (Fresh wt. in gm. at 'X' DAS / Fresh wt. in gm. at 10 DAS) x 100

Where, X = Actual Days after sowing

References

Andersen, T.M., Polle, E. and Konzak, C.F. (1987) Screening spring wheat for drought tolerance; *In* Genetic aspects of plant mineral nutrition, H.W. Gabelman and B.C. Loughman (Eds.). pp 79-86.

Blum, A., Sinmena, B. and Ziv, O. (1980) An evaluation of seed and seedling drought tolerance screening tests in wheat. *Euphytica,* 29: 727-736.

Clark, J.M. and Mc Craig, T.N. (1982) Evaluation of techniques for screening for drought resistance in wheat. *Crop Sci.,* 22: 503-506.

Kahovec, J. Fox, R.B. and Hatada, K. (2002) "Nomenclature of regular single-strand organic polymers". *Pure and Applied Chemistry.* 74 (10): 1921–1956.

Singh, P.N., Prasad, R., Salim, M. and Sharg, A. (1984). An improved system for subjecting plants to water stress. *Biol. Plant.* 26: 16-21.

Student Activities

Activity 1: Laboratory Exercise

From the given sample of different varieties identify the tolerant and susceptible varieties against drought by PEG method.

Activity 2: Answer these questions

Q.1. What is full form of PEG, PEO and POE?

Q.2. What is empirical chemical formula of PEG?

Q.3. Which property of PEG makes it suitable to produce high osmotic pressure?

Q.4. How many days after sowing (DAS) the PEG treatment should start?

Q.5. For experiments how much percent of PEG solution need to be prepared?

Q.6. When the last observation should take in this experiment?

2

Identification of Salt Tolerant Crop Variety by Glycine Betaine (GB) Method

Objective: To identify salt tolerant genotypes by Glycine Betaine (GB) method in laboratory.

Background information:

Salt or salinity is a biggest danger that causes reduction in crop productivity worldwide. Salinity reduces the germination percentage, plant vigour and ultimately loss in final crop yield. There are different physiological parameters that are affected by high level of salinity *viz.,* oxidative stress, physical alteration of bio-membrane, nutritional imbalance, loss of cell division & cell expansion, water stress, genotoxicity and ion toxicity. Together, these effects reduce plant growth, development and survival. During the onset and development of salt stress within a plant, all the major processes such as photosynthesis, protein synthesis and energy and lipid metabolism are affected. During initial exposure to salinity, plants experience water stress, which in turn reduces leaf expansion (Petronia, *et al.* 2011).

The water-soluble compounds which, are low in molecular weight, accumulate at certain concentration known as "compatible solutes" or "osmolytes" is the common strategy adopted by many organisms to combat the abiotic stresses. Some common compatible solutes are betaines, sugars (mannitol, sorbitol and trehalose), polyols, polyamines and amino acid (proline). Their accumulation is favored under water-deficit or salt stress as they provide stress tolerance to cell without interfering cellular machinery. The tolerant or sensitive species show difference in the stress tolerance level based on levels of accumulation of these compounds during abiotic stress condition. Accumulation of osmolytes like glycinebetaine (GB) in cells is known to protect organisms against abiotic stresses via osmoregulation or osmoprotection. Other roles of GB like cellular macromolecule protection and ROS detoxification have been suggested as mechanisms responsible for abiotic stress tolerance. In addition, GB also influences expression of several endogenous genes in plants.

Materials

1. Two known salt tolerant and susceptible cotton genotypes
   a. Genotype A – Cotton with low salt tolerance (Salt susceptible; SS)
   b. Genotype B – Cotton with high salt tolerance (Salt tolerant; ST)
2. Liquid nitrogen
3. Mortar-pestle
4. Deionized water
5. Filter paper and funnel
6. 2N H_2SO_4 and 1N H_2SO_4
7. Eppendr of tubes
8. KI- I_2 reagent
9. Vortex mixer
10. Refrigerator
11. High speed Cooling Centrifuge machine
12. 1, 2- Dichloroethane
13. UV-Vis Spectrophotometer
14. Glycine Betaine standard
15. Hoagland solution
16. NaCl, etc.

Procedure

Salinity treatment to the plant and sampling

1. Propagate the cotton seeds on the surface of sand (sand should be washed several times with deionised water to leach out the salts and autoclaved) and cover with 2 cm. layer of sand followed by keeping under the controlled environmental condition (25/10 °C day/night temperatures) in green house.
2. Regularly irrigate to the plants with half-strength Hoagland solution to avoid any physiological stress and to maintain high relative humidity condition.
3. Treat three week old seedlings with salt (NaCl) concentration of 0 mM (control), 50 mM, 100 mM and 200 mM.

4. Take samples from the shoots at 0, 24, 48 and 72 hours after treatment (hat).

Estimation of Glycine Betaine should be done by following the method of Grieve and Grattam, (1983)

1. Grind the leaf tissues (2.0g) to a fine powder using liquid nitrogen in mortar-pestle.
2. Shake the finely ground 0.5g plant sample from each genotype mechanically with 20 ml of deionized H_2O for 24 hours at 25 °C.
3. Filter the samples and make the dilution of extract in 1:1 ratio with 2N H_2SO_4.
4. Measure the aliquots (0.50 ml) into 2.0 ml eppendrof tubes and cool in ice-water for 1 hr.
5. Add cold KI- I_2 reagent (0.20 ml) to each tube and gently stir the reactants on a vortex mixer.
6. Store the tubes at 0-4 °C for 16 hrs and then centrifuge at 10000 rpm for 15 mins at 0 °C.
7. Aspirate the supernatents carefully and dissolve. Dissolve per-iodide crystals in 9.0 ml of 1, 2- dichloroethane.
8. Perform vigorous vortex mixing frequently for complete solubilisation in the developing solvent.
9. After 2-2.5 hrs, measure the absorbance at 365 nm with a spectrophotometer.
10. Prepare reference standard of GB (50-200 μg/ml) in 1N H_2SO_4. Prepare standard curve and calculate the GB content of sample.

Observation

Genotypes	NaCl Conc.	Glycine betaine content (μmol g^{-1} FW) Treatment duration			
		0 hat	24 hat	48 hat	72 hat
Genotype A (SS)	0 mM	____	____	____	____
	50 mM	____	____	____	____
	100 mM	____	____	____	____
	200 mM	____	____	____	____
Genotype B (ST)	0 mM	____	____	____	____
	50 mM	____	____	____	____
	100 mM	____	____	____	____
	200 mM	____	____	____	____

References

Grieve, C.M. and Grattan, S.R. (1983). Rapid assay for the determination of water soluble quaternary ammonium compounds. *Plant Soil,* 70: 303-307.

Petronia, C., Maria, G.A., Giovanni, P., Amodio, F. and Pasqualina, W. (2011) Salinity Stress and Salt Tolerance: *In* Abiotic Stress in Plants - Mechanisms and Adaptations, Shanker, A. and Venkateswarl, B. (Eds.), Chapter 2.

Student Activities

Activity 1: Laboratory exercise

Identify salt tolerant genotypes by Glycine Betaine (GB) method.

Activity 2: Answer these questions

Q.1. What is adverse effect of salt on plant growth?

Q.2. Which metabolic processes are affected by salt stress?

Q.3. Enlist the name of some osmolytes.

Q.4. Why reduction in leaf expansion is observed during salinity stress?

Q.5. Explain the role of osmolyte in imparting tolerance against abiotic stress?

Q.6. How glycinebetaine (GB) protect to the plant against salt stress?

3

Total Protein Estimation by Folin-Lowry's Method

Objective: To estimate total protein in the plant sample by Folin-Lowry's method.

Background information:

Protein assay can be done to determine the level of protein in solution by the Folin-Lowry biochemical method. The concentration of protein leads to change in color of the solution and based on different color in different protein concentration solution the change in color may be digitized by colorimeter or spectrophotometer. This technique was developed by Oliver H. Lowry in the year 1940s. The tyrosine and trytophan amino acid residues in a protein sample produces purple or blue color in the absorption maxima at 660 nm by using Folin-Ciocalteau (having phosphate and sodium tungstate molybdate) chemical reagent. The color intensity of protein sample depends on the concentration of above aromatic amino acids (tyrosine and tryptophan), hence every sample will vary for color. For protein estimation the BSA (Bovine serum albumin) is widely used as standard protein molecule due to its cost effectiveness, purity and easy availability. This method of protein estimation is very sensitive and can detect up to 10 μg/ml. This is a relative method of protein estimation and may be interfere by salts, Tris buffer, nonionic & cationic detergents, EDTA, lipids and carbohydrates. For reproducible assay the incubation time, pH (9-10.5) is critical in this method (Lowry, *et al.* 1951).

Materials

1. BSA stock solution (1mg/ml),
2. Analytical reagents:
   (a) Mix 50 ml sodium carbonate (2%) in 50 ml 0.1N NaOH.
   (b) Mix 10 ml copper sulphate (1.56%) in 10 ml sodium potassium tartarate (2.37%).

(c) In order to make ready analytical reagent mix 2 ml solution b with 100 ml solution a.

3. In order to make working 1N Folin-Ciocalteau reagent the commercial Folin-Ciocalteau 2N reagent should be diluted with equal amount of distilled water just prior to use on the day of experiment (2 ml distilled water + 2 ml 2N commercial Folin-Ciocalteau reagent).

Procedure

1. Prepare different dilutions of Bovine serum albumin protein solution through mixing stock (the concentration of stock BSA is 1 mg/ml) with water as per written in the given table. In every test tube the volume must be 5 ml. The range of BSA protein will be 0.05 mg/ml to 1 mg/ml.
2. From the above prepared dilutions, take 0.2 ml BSA solution in every test tube and further add 2 ml analytical reagent (alkaline copper sulphate). The solution should be mixed well on vortex mixture.
3. Incubate the above solution for 10 minutes at ambient temperature.
4. Add 0.2 ml Folin-Ciocalteau reagent in each test tube and keep for incubation at ambient temperature for 30 minutes. Take readings of color change by colorimeter or spectrophotometer at 660 nm wavelength with a blank.
5. The OD value should be plotted on the graph against BSA concentration to the absorbance and standard calibration curve must be prepared.
6. Based on standard calibration curve and OD value of unknown sample the protein concentration of unknown sample may be calculated using given formula.

Observation

Sr. No.	BSA (ml)	Water (ml)	Sample conc. (mg/ml)	Sample vol. (ml)	Alk. $CuSO_4$ (ml)	Lowry reagent (ml)	OD (660 nm)
1.	0.25	4.75	0.05	0.2	2	0.2	____
2.	0.05	4.5	0.1	0.2	2	0.2	____
3.	1	4	0.2	0.2	2	0.2	____
4.	2	3	0.4	0.2	2	0.2	____
5.	3	2	0.6	0.2	2	0.2	____
6.	4	1	0.8	0.2	2	0.2	____
7.	5	0	1.0	0.2	2	0.2	____

Calculation

Find out the protein content of unknown sample by dilution factor and standard graph.

Calculation of Dilution Factor (DF):

Calculation of Graph Factor (GF):

Protein content = OD X GF X DF

References

Lowry, O.H., Rosebrough, N.J., Farr, A.L., and Randall, R.J. (1951) *J.Biol.Chem* **193**: 265 (The original method).

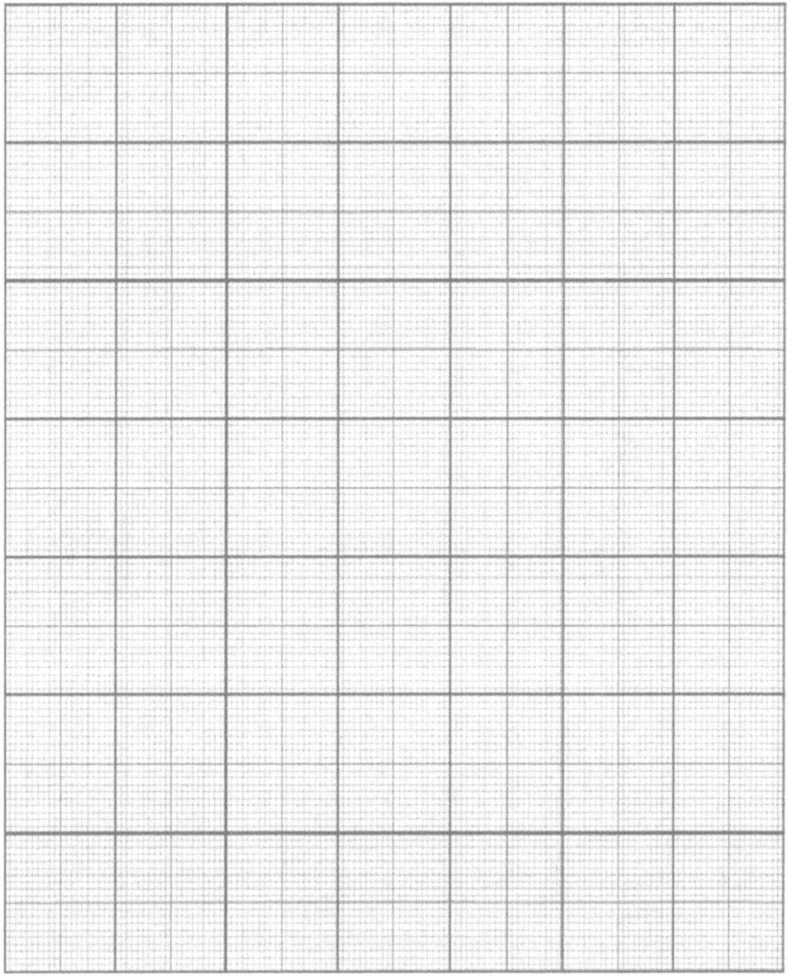

Graph

Student Activities

Activity 1: Laboratory exercise

Estimate total protein in given plant sample by Folin-Lowry's method.

Activity 2:Answer these questions

Q.1. Which amino acid in the protein produces blue color at 660 nm during Folin-Lowry method of protein estimation?

Q.2. What are the chemical constituents of Folin- Ciocalteau reagent?

Q.3. What is BSA? Write down full form of BSA.

Q.4. In which pH range the Folin-Lowry method of protein estimation works?

Q.5. Which compounds may interfere in the Folin-Lowry method of protein assay?

Q.6. When the method of Folin-Lowry was developed?

4

Protein Electrophoresis Using SDS-PAGE

Objective: To separate protein fragments by using SDS-PAGE method.

Background information

Electrophoresis is the motion of dispersed particles relative to a fluid under the effect of limited uniform field of electricity (Dukhin, and Derjaguin, 1974). Ferdin and Frederic Reuss (Reuss, 1809) first time observed such phenomena of electrokinetic nature in 1807. They found that the clay particles move in the water by giving constant supply of current. It happened due to interface of charged particles surface and the surrounding media. The technique is used widely in chemistry, biochemistry and molecular biology to separate different molecules based on difference in charge, binding affinity and size. Electrophoresis may be of two types *viz.*, Anaphoresis and Cataphoresis. Anaphoresis is the electrophoresis of anions (-ve charged ions) while, Cataphoresis is electrophoresis of cations (+ve charged ions). Electrophoresis is a major analytical technique in biochemistry that is used for separation of macromolecules like protein, enzymes and nucleic acids. Electrophoresis is a technique used in laboratories in order to separate macromolecules based on size.

PAGE is abbreviated for Polyacrylamide gel electrophoresis that describes a technique widely used in biochemistry, forensics, genetics, molecular biology and biotechnology to separate biological macromolecules, usually proteins or nucleic acids, according to their electrophoretic mobility. Mobility is a function of the length, conformation and charge of the molecule.

For proteins, sodium dodecyl sulfate (SDS) or sodium lauryl sulfate (SLS) is an anionic detergent that denatures secondary and non–disulfide–linked tertiary structures, and apply net negative charge to each peptide proportionate to its mass and thus linearized proteins (Laemmli, 1970). The disulfide bonds found in the protein complexes can be disrupt by β- Mercaptoethanol, which helps to further linearize the protein. This procedure is called SDS-PAGE. In most proteins, the binding of SDS to the polypeptide chain imparts an even distribution of charge per unit mass, thereby resulting in a fractionation by approximate size during electrophoresis (Rath *et al*., 2009). Polyacrylamide gel (PAG) may be made in different pore sizes and it is a synthetic, thermo-stable, transparent, strong, chemically relatively inert gel (Rüchel *et al.* 1978). Three factors determine gel pore size and its reproducibility; the total amount of acrylamide present (%T) (T = Total concentration of bisacrylamide and acrylamide), the amount of cross-linker (%C) (C = bisacrylamide concentration), and the time of polymerization of acrylamide

Materials

(a) Stock acrylamide soluion (30%)

29.2 g Acrylamide

0.8 g N,N' – Methylene bis acrylamide

Final volume made up to 100 ml using H_2O.

(b) Stock 1.5 M Tris-HCl

Dissolve tris-buffer (18.16g) in 60 ml distilled water, adjust the pH to 8.8 with HCl and make final volume up to 100 ml using H_2O.

(c) Stock 0.5 M Tris-HCl

Dissolve tris-buffer (3 g, 0.025 M) in 35 ml H_2O, adjust the pH to 6.8 and make final volume up to 50 ml using H_2O.

(d) Electrode buffer (pH 8.3)

Dissolve tris-buffer (3 g, 0.025 M) and 14.4 g Glycine (0.192 M) in H_2O and make up the final volume 1 litre. To perform SDS-PAGE 10 ml of 10 % SDS should be added and final volume should be adjusted to 1 litre.

(e) 10 % APS (Ammonium per sulfate): Dissolve 100 mg APS in 1 ml H_2O. Freshly prepared during gel preparation.

(f) TEMED (N,N,N'N' – Tetra methylethelendiamine).

(g) 10 % SDS: Dissolve 10g SDS in 100 ml H_2O.

(h) Gel loading dye

50 mM Tris-HCl (pH 6.8)

100 mM DTT (Dithiothreitol)

SDS (2%) (only in case of SDS-PAGE)

BPB (0.1%) (Bromophenol blue)

Glycerol (10%)

(i) Staining dye: (for protein)

Dissolve 0.1 g coomassie blue R-250 in 100 ml methanol solution (40): acetic acid (10): distilled water (50).

(j) Prepare destaining solution by adding methanol, acetic acid and H_2O in 40:10:50 ratio.

(k) Fix the gel with 7% acetic acid.

(l) Protein extraction buffer:

Tris-HCl (0.2 M, pH 7.2)

2% Triton X-100

1Mm EDTA (pH 8.0)

1% PVPP

1% β-mercaptoethanol (β-ME)

(m) Preparation of 12 % running gel (for protein)!

3.4 ml Double distilled ionized water

4.0 ml 30% Degassed Acrylamide/Bis

2.5 ml 1.5 M Tris-HCl, pH 8.8

0.1 ml 10% w/v SDS

50 µl 10% APS

5 µl TEMED

! Formulation for 10 ml

(n) Preparation of 4 % staking gel !

6.1 ml Double distilled ionized water

1.3 ml 30% Degassed Acrylamide/Bis

2.5 ml 0.5 M Tris-HCl, pH 6.8

0.1 ml 10% w/v SDS

50 µl 10% APS

10 µl TEMED

! Formulation for 10 ml.

(o) Vertical Gel electrophoresis unit, etc.

Procedure

1. Homogenize 1 g leaf tissues with 1 ml of protein extraction buffer.
2. Centrifuge the homogenates at 15,000 g for 15 minutes at 4 °C.
3. Mix the gel loading dye containing SDS with clear supernatant containing 50 µg proteins and load on to the gel.
4. Conduct electrophoresis on vertical SDS-PAGE (12%) at 30 mA for 3-4 hours.
5. Wash the gel to remove excess of SDS.
6. Stain the gel for proteins with coomassie blue stain for 1-2 hours.
7. Destain the gel with destaining solution with 3-4 changes after every 20 minutes.
8. Fix the destained gel with 7% acetic acid.

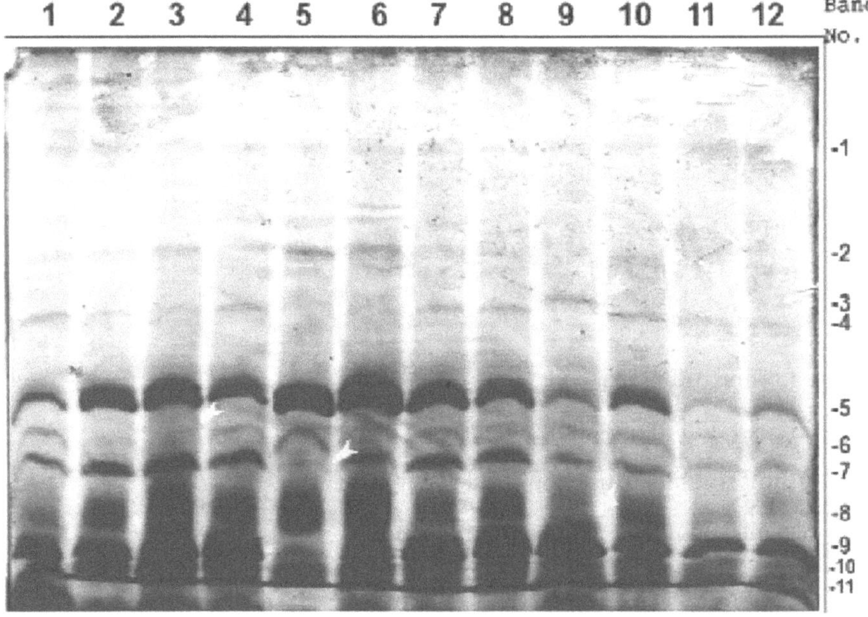

Fig 1: SDS-PAGE electrophoregrams of seed proteins of chilli (*Capsicum annuum* L.). (Source: Singh, *et al.* 2009)

Observation: (A representative SDS-PAGE gel stained with commassie blue is shown in the fig. 1 for example):

References

Dukhin, S.S. and Derjaguin, B.V. (1974). Electrokinetic Phenomena. J. Willey and Sons.

Laemmli, U.K. (1970). "Cleavage of structural proteins during the assembly of the head of bacteriophage T4". *Nature.* 227(5259): 680–685. *doi:10.1038/227680a0*

Rath, A., Glibowicka, M., Nadeau, V.G., Chen, G. and Deber, C.M. (2009). "Detergent binding explains anomalous SDS-PAGE migration of membrane proteins". *Proceedings of the National Academy of Sciences.* 106(6): 1760–1765. doi:10.1073/pnas.0813167106.

Reuss, F.F. (1809). "*Sur un nouvel effet de l'électricité galvanique*". *Mémoires de la Société Impériale des Naturalistes de Moscou.* 2: 327–337.

Rüchel, R., Steere, R.L. and Erbe, E.F. (1978). "Transmission-electron microscopic observations of freeze-etched polyacrylamide gels". *J Chromatogr.* 166(2): 563–575. doi:10.1016/S0021-9673(00)95641-3

Singh, D., Mehta, R. and Talati, J.G. (2009). Morphological, biochemical and electrophoretic evaluation of chilli (*Capsicum annum* L.) genotypes. *Indian J Agric Biochem*, 22(2):73-77.

Student Activities

Activity 1: Laboratory exercise

A. Extract and separate protein fragments by SDS-PAGE method.

Activity 2: Answer these questions

Q.1. Define electrophoresis?

Q.2. What is PAG?

Q.3. What is SDS or SLS?.

Q.4. What is PAGE?

Q.5. Which factors determine the pore size of a gel and the reproducibility in gel pore size?

Q.6. Write down the importance of SDS in protein electrophoresis?

5

Enzyme Extraction and Assay Study of Polyphenol Oxidase (PPO; EC 1.14.18.1)

Objective: To extract and perform enzyme assay of polyphenol oxidase from plant.

Background information:

Polyphenol oxidase (PPO or monophenol monooxygenase or polyphenol oxidase I, chloroplastic) is a tetramer that contains four atoms of copper per molecule, and binding sites for two aromatic compounds and oxygen (www.worthington-biochem.com). The enzyme catalyses the *o*-hydroxylation of monophenol molecules in which the benzene ring contains a single hydroxyl substituent to *o*-diphenols (phenol molecules containing two hydroxyl substituents). It can also further catalyse the oxidation of *o*-diphenols to produce *o*-quinones.

PPO causes the rapid polymerization of *o*-quinones to produce black, brown or red pigments (polyphenols) that cause fruit browning. The amino acid tyrosine contains a single phenolic ring that may be oxidised by the action of PPOs to form *o*-quinone. Hence, PPOs are some time referred as tyrosinases (Mayer, 2006).

Jolley *et al.* (1974) referred to PPO as 4 electron-transferring phenol oxidase and oxygen. It is responsible for browning reactions throughout the phylogenetic scale.

Materials and treatment condition

As described in practical no. 2 (salt treated plant samples) and as follows:

1. 0.2 M potassium phosphate buffer, pH 6.8 (Extraction buffer)
2. 0.050 M catechol (Substrate), etc.

Enzyme extraction

1. Grind 300 mg of leaves in 3 ml of 0.2 M potassium phosphate buffer, pH 6.8.
2. Centrifuge the homogenate at 10,000 rpm for 15 min at 4 °C.
3. Use the supernatant for enzyme assay.

Enzyme Assay

1. The reaction mixture contains 2.9 ml of catechol (0.050 M catechol in 0.2 M phosphate buffer pH 6.8).
2. Initiate the reaction by addition of 0.1 ml of enzyme extract.
3. Read the color change by the oxidized catechol at 490 nm for three minute at an interval of 30 second.

Calculation

Express the enzyme activity as change in O.D./min/g protein (Thimmaiah, S.K., 2006).

Observation

Genotypes	NaCl Conc.	PPO enzyme activity (O.D./min/g protein) Treatment duration			
		0 hat	24 hat	48 hat	72 hat
Genotype A (SS)	0 mM	——	——	——	——
	50 mM	——	——	——	——
	100 mM	——	——	——	——
	200 mM	——	——	——	——
Genotype B (ST)	0 mM	——	——	——	——
	50 mM	——	——	——	——
	100 mM	——	——	——	——
	200 mM	——	——	——	——

References

Jolley, R., Evans, L., Makino, N. and Mason, H. (1974): Oxytyrosinase , *J Biol Chem* 249, 335,

Mayer, A.M. (2006). "Polyphenol oxidases in plants and fungi: Going places? A review". *Phytochemistry*. 67(21):2318–233.doi:10.1016/j.phytochem. PMID16973188

Thimmaiah, S.K. (2006) Standard Methods of Biochemical Analysis. Publisher, Kalyani Publishers, ISBN, 8176630675

www.worthington-biochem.com

Student Activities

Activity 1: Laboratory exercise

Extract PPO enzyme from given treated sample and do enzyme assay.

Activity 2: Answer these questions

Q.1. What are other names of polyphenol oxidase?

Q.2. Which metal compound and binding sites are found in PPO and in how many numbers?

Q.3. Which type of enzyme catalysis reaction occurs with PPO enzyme?

Q.4. How PPO leads to browning in fruits?

Q.5. Which extraction buffer is needed for PPO enzyme extraction?

Q.6. What is the substrate for PPO enzyme?

6

Enzyme Extraction and Assay Study of Catalase (EC 1.11.1.6)

Objective: To extract and perform enzyme assay of catalase from plant.

Background information

Catalase is a commonly distributed enzyme and present in approximately all organisms exposed to oxygen (such as bacteria, plants and animals). Catalysis of H_2O_2 into H_2O and O_2 is done by catalase enzyme (Goodsell, 2004).

$$H_2O_2 \rightarrow 2\ H_2O + O_2$$

It is a very important enzyme in protecting the cell from oxidative damage by ROS (reactive oxygen species). Similarly, the highest turnover number among all enzymes is found in catalase; one molecule of catalase enzyme can change millions of H_2O_2 molecules to H_2O and O_2 per second (Boon, *et al.* 2007).

Four polypeptide chains are found in catalase enzyme, so it may be called as a tetramer, The length of each polypeptide chain is more than 500 amino acids (Maehly and Chance, 1954). It can react with H_2O_2 because of four porphyrin heme (iron) groups. Optimum pH for human catalase is approximately 7 (Aebi, 1984), and has a fairly broad maximum (the rate of reaction does not change appreciably at pHs between 6.8 and 7.5) (BRENDA, 2009). The pH optimum for other catalases varies between 4 to 11 based on different species (Toner, *et al.* 2000). Optimum temperature for the activity of catalase enzyme may also vary from species to species.

The test of catalase in a microbial or tissue sample can be tested by adding H_2O_2 and observing the reaction. The formation of bubbles, oxygen, indicates a positive result. This easy assay, which can be seen with the naked eye, without the aid of instruments, is possible because catalase has a very high specific activity, which produces a detectable response. Alternative splicing may result in different protein variants (Chelikani, *et al.,* 2004).

Materials and treatment condition

As described in Practical no. 2 (salt treated plant samples) and as follows:

1. 0.2 M potassium phosphate buffer, pH 6.8 (Extraction buffer)
2. 0.03% H_2O_2 (Substrate), etc.

Enzyme extraction

1. Grind 300 mg of leaves in 3 ml of 0.2 M potassium phosphate buffer, pH 6.8.
2. Centrifuge the homogenate at 10,000 rpm for 15 min at 4 °C.
3. Use the supernatant for enzyme assay.

Enzyme Assay

1. Assay the catalase (EC 1.11.1.6) activity (Aebi, 1984) in a reaction mixture containing 50 mM phosphate buffer, pH 7.0, 0.03% H_2O_2 and 0.1 ml enzyme extract.
2. Record the decomposition of H_2O_2 at 240 nm.

Calculation

Express the enzyme activity as change in O.D./min/g protein.

Observation

Genotypes	NaCl Conc.	Catalase enzyme activity (O.D./min/g protein) Treatment duration			
		0 hat	24 hat	48 hat	72 hat
Genotype A (SS)	0 mM	——	——	——	——
	50 mM	——	——	——	——
	100 mM	——	——	——	——
	200 mM	——	——	——	——
Genotype B (ST)	0 mM	——	——	——	——
	50 mM	——	——	——	——
	100 mM	——	——	——	——
	200 mM	——	——	——	——

References

Aebi, H. (1984). "Catalase *in vitro*". Methods in Enzymology. 105: 121–6. doi:10.1016/S0076-6879(84)05016-3. ISBN 978-0-12-182005-3. PMID 6727660.

Boon, E.M., Downs, A. and Marcey D. (2007) "Catalase: H_2O_2: H_2O_2 Oxidoreductase". Catalase Structural Tutorial Text.

Brenda (2009). "EC 1.11.1.6 - catalase": The Comprehensive Enzyme Information System. Department of Bioinformatics and Biochemistry, Technische Universität Braunschweig.

Chelikani, P., Fita, I. and Loewen, P.C. (2004). "Diversity of structures and properties among catalases". *Cellular and Molecular Life Sciences*. 61(2): 192–208. doi:10.1007/s00018-003-3206-5. PMID 14745498.

Goodsell, D.S. (2004). "Catalase". Molecule of the Month. RCSB Protein Data Bank.

Maehly, A.C. and Chance, B. (1954). "The assay of catalases and peroxidases". Methods of Biochemical Analysis. Methods of Biochemical Analysis. 1: 357–424. doi:10.1002/9780470110171.ch14. ISBN 978-0-470-11017-1. PMID 13193536.

Toner, K., Sojka, G. and Ellis, R. (2000) "A Quantitative Enzyme Study; CATALASE". bucknell.edu.

Student Activities

Activity 1: Laboratory exercise

A. Extract catalase enzyme from given treated sample and do enzyme assay.

Activity 2: Answer these questions

Q.1. Write down the chemical reaction takes place by catalase enzyme?

Q.2. Write down the role of catalase enzyme in living system.

Q.3. How many polypeptide chains are present in the catalse enzyme and how many amino acids are found in each polypeptide chain?

Q.4. Write down the number of porphyrin heme (iron) groups that are present in the catalase enzyme.

Q.5. Write down the optimum pH for human catalase enzyme.

Q.6. What is the turn over number of catalase enzyme?

7

Enzyme Extraction and Assay Study of Guaiacol Peroxidase (POX) (EC 1.11.1.7)

Objective: To extract and perform enzyme assay of guaiacol peroxidase from plant.

Background information:

Peroxidases are a large family of enzymes that typically catalyze a reaction of the form:

$$ROOR' + 2e^- + 2H^+ \xrightarrow{\text{Peroxidase}} ROH + R'OH$$

Where, $2e^-$ comes from electron donor

Peroxidase (POX) is an oxidoreductase enzyme (EC 1.11.1.7) that catalyses the oxidation of a wide variety of organic and inorganic substrates using hydrogen peroxide as the electron acceptor (Banci, 1997). POXs are ubiquitous in plants and have various physiological roles in plant tissues, including oxidation of biomolecules by accumulation of active oxygen, like superoxide radical ($O_2^{\cdot-}$), the hydroxyl radical (OH) and H_2O_2 (Scialabba, *et al.* 2002). In plants, POX plays different roles, including construction, rigidification and eventual lignifications of cell walls. The enzyme is able to cross-link phenolic residues of cell wall polysaccharides and glycoproteins, serving to strengthen the cell wall components (Lee, *et al.* 2007; Kim and Lee, 2005). POXs also protect the tissue from damages and infections caused by pathogenic microorganisms. This wound-healing response appears after POX induction by stress, caused by for example, high salt concentration, physical wounding or microorganisms (Gulcin and Yildirim, 2005; Sisecioglu *et al.* 2010).

Other names of peroxidases are lactoperoxidase; guaiacol peroxidase; plant peroxidase; Japanese radish peroxidase; horseradish peroxidase (HRP); soybean peroxidase (SBP); extensin peroxidase; heme peroxidase; oxyperoxidase;

protoheme peroxidase; pyrocatechol peroxidase; scopoletin peroxidase, *Coprinus cinereus* peroxidase, *Arthromyces ramosus* peroxidase (Paul, 1963).

Materials and treatment condition

As described in practical no. 2 (salt treated plant samples) and as follows:

1. 50 mM sodium phosphate buffer (pH 7.4)
2. Guaiacol (Substrate), etc.

Enzyme extraction

1. Prepare enzyme extract for determination of guaiacol peroxidase (POX) activities from 0.3 g of cotton leaf by homogenizing with a pre-chilled mortar and pestle under ice cold condition in 3 ml of extraction buffer, containing 50 mM sodium phosphate buffer (pH 7.4) with the addition of 1 mM EDTA and 1% (w/v) polyvinylpyrolidone (PVP).
2. Centrifuge the homogenates at 10,000 rpm for 20 minutes at 4 °C and keep the supernatant for further enzyme assay (Costa, *et al.*, 2002).

Enzyme Assay

1. Determine the activity of POX enzyme in the homogenates by observing the increase in absorption at 470 nm by formation of tetraguaiacol (=26.6 mM^{-1} cm^{-1}) in a reaction mixture containing 50 mM sodium phosphate buffer pH 7.0, 0.1 mM EDTA, 0.05 ml enzyme extract, 10 mM guaiacol and 10 mM H_2O_2 (Costa, *et al.* 2002).

   4 Guaiacol + 4 H_2O_2 POX Tetraguaiacol + $8H_2O$

   →

Calculation

Express the enzyme activity as change in O.D./min/g protein.

Observation

Genotypes	NaCl Conc.	Guaiacol peroxidase enzyme activity (O.D./min/g protein) Treatment Duration			
		0 hat	24 hat	48 hat	72 hat
Genotype A (SS)	0 mM	____	____	____	____
	50 mM	____	____	____	____
	100 mM	____	____	____	____
	200 mM	____	____	____	____
Genotype B (ST)	0 mM	____	____	____	____
	50 mM	____	____	____	____
	100 mM	____	____	____	____
	200 mM	____	____	____	____

References

Banci, L. (1997) Structural properties of peroxidases. *J. Biotechnol.* 53:253–63.

Costa, M.L., Civello, P.M., Chaves, A.R. and Martinez, G.A. (2002). Effect of ethephon and 6-benzylaminopurine on chlorophyll degrading enzymes and a peroxidase-linked chlorophyll bleaching during post-harvest senescence of broccoli (*Brassica oleracea* L.) at 20°C. *Postharvest Biology and Technology*, 35:191–199.

Gulcin, I. and Yildirim, A. (2005) Purification and characterization of peroxidase from *Brassica oleracea* var. Acephala. *Asian J. Chem.* 17:2175–2183.

Kim, S.S. and Lee, D.J. (2005) Purification and characterization of a cationic peroxidase Cs in *Raphanus sativus*. *J. Plant Physiol.* 162: 609–617.

Lee, B.R., Kim, K.Y., Jung, W.J., Avice, J.C., Ourry, A. and Kim, T.H. (2007) Peroxidases and lignification in relation to the intensity of water deficit stress in white clover (*Trifolium repens* L.). *J. Exp. Bot.* 58:1271–1279.

Paul, K.G (1963) Peroxidases. *In*: Boyer, P.D., Lardy, H. and Myrbäck, K. (Eds), The Enzymes, 2nd edn, vol. 8, Academic Press, New York, pp. 227-274.

Scialabba, A., Bellani, L.M. and Dell'Aquila, A. (2002) Effects of ageing on peroxidase activity and localization in radish (*Raphanus sativus* L.) seeds. *Eur. J. Histochem.* 46:351–358.

Sisecioglu, M., Gulcin, I., Cankaya, M., Atasever, A., Sehitoglu, M.H., Kaya, H.B. and Ozdemir, H. (2010) Purification and characterization of peroxidase from Turkish black radish (*Raphanus sativus* L.). *J. Med. Plants Res.* 4:1187–1196.Paul, K.G (1963) Peroxidases. *In*: Boyer, P.D., Lardy, H. and Myrbäck, K. (Eds), The Enzymes, 2nd edn, vol. 8, Academic Press, New York, pp. 227-274.

Student Activities

Activity 1: Laboratory exercise

A. Extract Guaiacol peroxidase enzyme from given treated sample and do enzyme assay.

Activity 2: Answer these questions

Q.1. Which chemical reaction occurs by guaiacol peroxidase enzyme?

Q.2. What is the role of guaiacol peroxidase enzyme in plant tissues?

Q.3. How guaiacol peroxidase enzyme strengthen plant cell wall?

Q.4. What are other names of peroxidase enzyme?

Q.5. Write extraction buffer used for guaiacol peroxidase enzyme?

Q.6. What is the substrate for guaiacol peroxidase enzyme and which molecule act as electron acceptor?

8

Enzyme Extraction and Activity Study of Superoxide Dismutase (EC 1.15.1.1)

Objective: To extract and perform enzyme assay of Superoxide dismutase from plant.

Background information:

Superoxide dismutase (SOD, EC 1.15.1.1) is the enzyme that alternately catalyzes the dismutation (or partitioning) of the superoxide (O_2^-) radical into either ordinary molecular oxygen (O_2) or hydrogen peroxide (H_2O_2). Superoxide is produced as a by-product of oxygen metabolism and, if not regulated, may cause cell damage (Hayyan, *et al.* 2016). Hydrogen peroxide is also responsible for cell damage and may be degraded by other enzymes like catalase. Thus, SOD is an important antioxidant defense in nearly all living cells exposed to oxygen. One exception is *Lactobacillus plantarum* and related lactobacilli, which use a different mechanism to prevent damage from reactive O_2^-.

The main function of SOD is like antioxidant in plants that protect from cell damage by ROS (Alscher, *et al.* 2002). Reactive oxygen species may be formed inside the living cell by several conditions like drought, excess pesticides & herbicides, mechanical injury, ozone, photoinhibition, high & low temperature, toxic metals, UV/gamma rays, nutrient deficiency and plant metabolic activity (Smirnoff, 1993; Raychaudhuri and Deng, 2008). To be specific, molecular O_2 is reduced to O_2^- (it is a ROS known as superoxide) by absorbing an excited electron produced from ETC.

Superoxide is responsible for cell damage by fragmenting DNA, oxidizing lipids and denaturing the enzymes. H_2O_2 and O_2 (less harmful reactants) is produced by the catalysis of SODs from superoxide (O_2^-).

SOD concentration increases with the increased level of oxidative stress condition. SOD is produced in different compartments of plants that helps in fighting oxidative stress more efficiently and effectively. There are three different classes of SOD metallic coenzymes that exist in plants. First, Fe SODs consist of two species, one homodimer (containing 1-2 g Fe) and one tetramer (containing 2-4 g Fe). It is probably most ancient SOD metalloenzymes and found within both prokaryotes and eukaryotes. Fe SODs are most abundantly localized inside plant chloroplasts, where they are indigenous. Second, Mn SODs consist of a homodimer and homotetramer species each containing a single Mn(III) atom per subunit. They are found predominantly in mitochondrion and peroxisomes. Third, Cu-Zn SODs have electrical properties that vary from other two classes. These are concentrated in the cytosol, chloroplast, and occasionally in extracellular space. Note that Cu-Zn SODs provide less protection than Fe SODs when present in chloroplast (Alscher, *et al.* 2002; Smirnoff, 1993; Raychaudhuri and Deng, 2008)

Materials and treatment condition

As described in practical no. 2 (salt treated plant samples) and as follows:

1. 50 mM phosphate buffer pH 7.8 (Extraction buffer)
2. Nitroblue Tetrazolium (NBT) (Substrate), etc.

Enzyme extraction

1. Determine the superoxide dismutase (EC 1.15.1.1) activity by measuring its ability to inhibit the photochemical reduction of nitroblue tetrazolium (NBT) using the method of Beauchamp and Fridovich (1971).

Enzyme Assay

1. Reaction mixture for SOD enzyme assay should contain 50 mM phosphate buffer pH 7.8, 13 mM methionine, 75 mM nitro blue tetrazolium (NBT), 2 mM riboflavin, 0.1 mM EDTA and 200 ml enzyme extract in a total volume of 3 ml.
2. Add riboflavin in the tubes at last, shake well and place below a light bank consisting of 15 W white fluorescent lamp.
3. Initiate the reaction by switching on the light and allow to react for 10 min.
4. Stop the reaction by switching of the lights and cover the tubes with a black cloth.
5. Read the absorbance at 560 nm.
6. A non-irradiated solution will not develop color and serve as a control.
7. The reaction mixture-lacking enzyme develops maximum color and color intensity decrease with decrease in volume of enzyme extract added.

Calculation

The unit SOD may be calculated as under:

$$1 \text{ unit of SOD} = \frac{\text{O.D. of control}}{\text{O.D. of sample}} - 1$$

Observation

Genotypes	NaCl Conc.	Superoxide Dismutase enzyme activity (O.D./min/g protein) Treatment duration			
		0 hat	24 hat	48 hat	72 hat
Genotype A (SS)	0 mM	——	——	——	——
	50 mM	——	——	——	——
	100 mM	——	——	——	——
	200 mM	——	——	——	——
Genotype B (ST)	0 mM	——	——	——	——
	50 mM	——	——	——	——
	100 mM	——	——	——	——
	200 mM	——	——	——	——

References

Alscher, R.G., Erturk, N. and Heath, L.S. (2002). "Role of superoxide dismutases (SODs) in controlling oxidative stress in plants". *Journal of Experimental Botany*. 53(372): 1331–41. doi:10.1093/jexbot/53.372.1331. PMID 11997379.

Beauchamp, C. and Fridovich, I. (1971). Superoxide dismutase: improved assays and an assay applicable to acrylamide gels. *Anal Biochem*. 44(1):276-87.

Hayyan, M., Hashim, M.A. and Al Nashef, I.M. (2016). "Superoxide Ion: Generation and Chemical Implications". Chem. Rev. 116 (5): 3029–3085. doi:10.1021/acs.chemrev.5b00407

Raychaudhuri, S.S. and Deng, X.W. (2008). "The Role of Superoxide Dismutase in Combating Oxidative Stress in Higher Plants". *The Botanical Review*. 66(1): 89–98. doi:10.1007/BF02857783.

Smirnoff, N. (1993). "Tansley Review No. 52 The role of active oxygen in the response of plants to water deficit and desiccation". *Plant Phytology*. 125.

Student Activities

Activity 1: Laboratory exercise

Extract SOD enzyme from given treated sample and do enzyme assay.

Activity 2: Answer these questions

Q.1. Which type of chemical reaction is generally catalyzed by superoxide dismutase enzyme?

Q.2. Explain the significance of SOD enzyme in living system.

Q.3. How ROS is formed in the plant system?

Q.4. How the stress can be counteract by SOD very effectively?

Q.5. How many known SOD metallic coenzymes exist in plants?

Q.6. What is the substrate for superoxide dismutase?

9

Peroxidase Isozyme Study by Electrophoresis Method

Objective: To study peroxidase isozyme variation in rice seedlings by electrophoresis.

Background information

For electrophoresis and PAGE please see practical no. 4. For peroxidase please see practical no. 7. (Page 17-22)

Isozymes (isoenzymes or more generally as multiple forms of enzymes) are the enzymes that are differing in the sequence of amino acids but they work on same substrate or in other words catalyze same biochemical reaction. The isozymes have difference in K_M values and regulatory properties. Isozymes are very useful for proper tuning of metabolic activities and to meet the specific need of particular tissue or developmental stage, e.g. lactate dehydrogenase (LDH). Isozymes are coded by homologous genes that have diversity during evolutionary process. Although, strictly speaking, allozymes represent enzymes from different alleles of the same gene, and isozymes represent enzymes from different genes that process or catalyse the same reaction, the two words are usually used interchangeably.

Materials

(a) Stock acrylamide soluion (30%)

29.2 g Acrylamide

0.8 g N,N' – Methylene bis acrylamide

Adjust volume to 100 ml with H_2O

(b) Stock 1.5 M Tris-HCl

Dissolve tris-buffer (18.16g) in 60 ml distilled water, then adjust pH to 8.8 with HCl and make volume to 100 ml with distilled water.

(c) Stock 0.5 M Tris-HCl

Dissolve tris-buffer (3 g, 0.025 M) in 35 ml H_2O, then adjust pH to 6.8 and make final volume up to 50 ml with H_2O.

(d) Electrode buffer (pH 8.3)

Dissolve tris-buffer (3 g, 0.025 M) and Glycine (14.4 g, 0.192 M) in H_2O and make volume to 1 litre.

(e) 10 % APS (Ammonium per sulfate): Dissolve 100 mg APS in 1 ml H_2O just prior to gel formation.

(f) TEMED (N,N,N'N' – Tetra methylethelendiamine).

(g) Gel loading dye

50 mM Tris-HCl (pH 6.8)

100 mM DTT (Dithiothreitol)

0.1% BPB (Bromophenol blue)

10% Glycerol

(h) Staining dye:

100 ml solution containing O-Dianisidine (1 g), acetic acid (9 ml) and 3% hydrogen peroxide [0.03% H_2O_2 was prepared in 0.01 M phosphate buffer (pH 6.0)]

(i) Prepare destaining solution by addition of methanol, acetic acid and H_2O in 40:10:50 ratio.

(j) Fix the gel with 7% acetic acid.

(k) Extraction buffer:

0.1 M phosphate buffer (pH 7.2) containing 1 mM polyvinyl pyrolidone (PVPP) and 10% Glycerol was used for.

(l) Preparation of 7% running gel (for isozymes)!

5.1 ml Double distilled ionized water

2.3 ml 30% Degassed Acrylamide/Bis

2.5 ml 1.5 M Tris-HCl, pH 8.8

50 μl 10% APS

5 μl TEMED

! Formulation for 10 ml

(m) Preparation of 4 % staking gel !

6.2 ml Double distilled ionized water

1.3 ml 30% Degassed Acrylamide/Bis

2.5 ml 0.5 M Tris-HCl, pH 6.8

50 µl 10% APS

10 µl TEMED

! Formulation for 10 ml.

(n) Vertical Gel electrophoresis unit, etc.

Procedure

1. Grind rice seedlings (1g) with glass powder in cold 0.1 M phosphate buffer (pH 7.2) containing 1 mM PVPP and 10% glycerol.
2. Centrifuge the homogenate at 10,000 g for 20 minutes at 4 °C and use the supernatant containing 50 ìg of protein for enzyme assay and electrophoresis.
3. Stain the gel for POX activity in the solution (100 ml) containing 1 g of O-Dianisidine, 9 ml of acetic acid and 3% hydrogen peroxide (Sadasivam and Manickam, 1992).

Observation: (A representative PAGE gel of peroxidase isozyme stained with 1% Orthodianisidine dye is shown in fig. 2 for example):

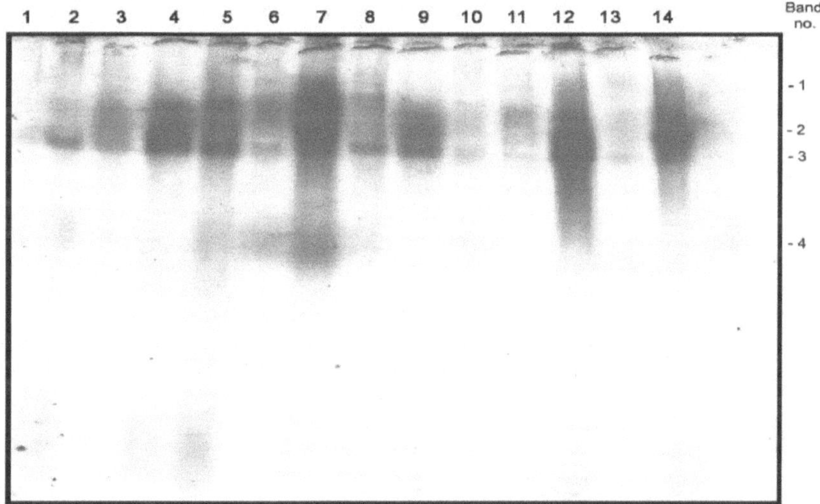

Fig. 2: Isozyme pattern for peroxidase in rice leaves at 5 DAG (Source: Singh, *et al.* 2009)

References

Sadasivam, S. and Manickam, A. (1992). Biochemical Methods for Agricultural Sciences. p 98-147.

Singh, D., Mehta, R. and Talati, J.G. (2009). Identification of rice (*Oryza sativa* L.) varieties through biochemical markers. *Indian J Agric Biochem*, 22(1):18-25.

Student Activities

Activity 1: Laboratory exercise

Study peroxidase isozyme variation in rice seedlings by electrophoresis.

Activity 2: Answer these questions

Q.1. Define isozyme giving example?

Q.2. How much gel percent should be prepared for isozyme study of perxoidase?

Q.3. How the peroxidase isozyme gel is stained after electrophoresis?

Q.4. What is homologous gene?

Q.5. What is allozyme?

Q.6. Which type of electrophoresis apparatus is required for isozyme electrophoresis?

10

DNA Isolation from Plant Sample and Quantification by Spectrophotometer

Objective: To isolate DNA from plant sample by CTAB method and quantification by spectrophotometer.

Background information:

DNA isolation from different plant tissues *viz.*, leaf, stem, root, flower and seed is a challenging task due to hindrance of variety of variety of biochemical compounds present and further problem increases with diverge plant species. In case of animal tissues the diversity in cellular bio-molecules in different species is much lower compared to plants. Two major phyto bio-molecules that have wide variation between species and pose serious problems in DNA isolation are: polyphenols and polysaccharides. These phyto bio-molecules if contaminated in DNA then may hinder further down applications like PCR, restriction digestion, sequencing, etc.

Several methods have been developed and reported that can remove polyphenols and polysaccharides efficiently during DNA isolation from plants. For this purpose two different chemicals are widely used, the first one is CTAB (cetyl trimethyl ammonium bromide) that is a cationic detergent and helps in removal of polysaccharides and the second one is polyvinylpyrrolidone that helps in removal of polyphenols. Due to above properties the CTAB based DNA extraction buffer is very common for plant DNA extraction.

Significance of different types of chemicals found in CTAB based plant DNA extraction method is explained as under:

CTAB

Plant DNA is highly polymerized and CTAB was established sometime ago as the best detergent to extract such DNA. The plant cell wall is also solubilized by this detergent in addition to solubilization of lipid membranes of internal organelles and denaturation of proteins (enzymes). Thus, the DNA is not hydrolyzed during the isolation process and as long as vortexing or vigorous shaking are avoided highly polymerized (i.e., very high intact) genomic DNA is obtained. CTAB is a powerful cationic surfactant, thus it disrupts membranes and releases DNA. Plant tissues are rich in complex polysaccharides and secondary metabolites like polyphenols, etc., which interfere and co-precipitate with DNA during the isolation procedure. CTAB along with some other chemical like PVP is applied to minimize the effect of these metabolites.

Chloroform

It solubilizes lipids and a lot of proteins to remove them from the DNA. Proteins and polysaccharides are separated from nucleic acids in the cell by using Chloroform. Chloroform bind with the complexed proteins and polysaccharides. Chloroform is denser than water solutions and thus after spinning this solution the chloroform and water will separate into two distinct phases. The lower phase will be chloroform that chemically attracts proteins and polysaccharides. The upper aqueous phase contains DNA. Plant tissues contain often more polysaccharides than animal tissues. If the polysaccharides are not removed properly then it may hinder several down applications like restriction digestion, etc.

Chilled isopropanol

DNA is less soluble in solutions containing isopropanol than in solutions containing ethanol. In contrast to precipitation with ethanol, which requires 23 volumes of alcohol, precipitation with isopropanol is performed with 60.7 volume of alcohol. Isopropanol is often the better choice when precipitating DNA from large volumes of solution. Precipitation with isopropanol should be done at ambient temperature to lessen the risk that solutes like sucrose or sodium chloride will be coprecipitated with the DNA. Isopropanol use is useful for precipitations where sample volume is large.

Chilled ethanol

DNA does not dissolve in ethanol. It precipitates the DNA - causes it to come out of solution. DNA is separated out of water-based solutions using chilled ethanol. This allows the DNA to be purified for subsequent genetic testing. Adding alcohol to a solution containing DNA is a simple method for obtaining pure DNA required and colder temperatures slow down enzymes that can break down DNA, giving better extraction results.

Deionized water

DNA get dissolve in water so we can store in deionized water for further use of DNA in the experiments. By using deionized water the interference of ions during DNA extraction may be nullified.

Materials

Components used for CTAB DNA extraction method

Sr.No.	Stocks	Final concentration	Volumes added in buffer
1.	Tris buffer stock, 1.0 M (pH 8.0)	100 mM	1.0 ml
2.	EDTA stock 0.5 M (pH 8.0)	50 mM	1.0 ml
3.	NaCl stock, 5 M	700 mM	2.8 ml
4.	β-Mercapto ethanol	1 %	100 μl
5.	CTAB (Cetyl Trimethyl Ammonium Bromide) stock	3%	300 mg
6.	PVP (Polyvinyl pyrrolidone)	1 %	100 mg
7.	Sterile millipore water	Volume 100 ml	

More chemicals for DNA extraction are as under:

i. Phenol:Chloroform:Isoamyl alcohol (25:24:1)

ii. Absolute alcohol

iii. Ethanol (70%)

iv. TE buffer (pH 8.0)

Preparation of reagents

1. 1 M Tris-HCl pH 8.0 (100 ml)

Dissolve 12.11 g tris base in of distilled water (80 ml). Adjust the pH to 8.0 with 1 N HCl make volume to 100 ml by H_2O. Autoclave and store at room temperature.

2. 0.5 M EDTA (100 ml)

Dissolve 18.612 g of Na_2EDTA. $2H_2O$ in distilled water (80 ml), adjust pH to 8.0 with NaOH (by adding 2 g of 10 N NaOH pellet). Stir vigorously for several minutes on a magnetic stirrer and adjust volume 100 ml.

(Na_2EDTA does not dissolve in water in absence of NaOH).

3. 5 N NaCl

Dissolve 19.62 g NaCl in distilled water (80 ml) and adjust volume to 100 ml with H_2O.

4. TE buffer (100ml)

1 M Tris buffer	1.0 ml
0.5 M EDTA	0.2 ml

Make volume to 100 ml. Autoclave and store at ambient temperature.

5. Phenol: Chloroform: Isoamyl alcohol (500 ml), etc.

Chloroform	240 ml
Re-distilled phenol	250 ml
Isoamyl alcohol	10 ml

Store in dark bottle at 4 °C.

Procedure

Extract total genomic DNA using CTAB (Cetyl Trimethyl Ammonium Bromide) method of Doyle and Doyle (1990) with few modifications. The steps are described as under with all necessary details:

DNA Extraction

1. Mix well all ingredients and warm CTAB extraction buffer (containing 700mM NaCl, 100mM Tris-HCl pH-8, 50mM EDTA pH-8, 1% PVP and 3% CTAB) at 65 °C till CTAB and PVP dissolve clearly.
2. Just before use, add 100µl of ß-mercaptoethanol (1%) and mix well.
3. Weigh 300 mg leaves of young seedling of seven days after sprouting. Etiolated leaves should be selected to avoid problem of pigments present in leaves.
4. Sterilize mortar and pestle with spirit or alcohol, dry and prechilled with liquid nitrogen. Crush the samples to get fine powder, and then scrape with spatula and fill in 2 ml eppendorf tubes.
5. Add immediately 1 ml hot extraction buffer and mix by vigorous shaking. Incubate the tubes at 65 °C for 1 hour. Shake the tubes thrice after each 15 min. to get high yield of DNA.
6. Mix same quantity of phenol: chloroform: isoamyl alcohol (25:24:1) in all tubes. Mix it by inverted shaking to form a white milky phase.

7. Centrifuge the tubes at 10,000 rpm at 15 °C for 10 min., because CTAB precipitates below 15 °C temperature.
8. Take the white transparent supernatant into fresh tube with micropipette tip. Avoid physical shaking to DNA at this stage.
9. Repeat above three steps twice.

DNA Precipitation:

1. Transfer the aqueous layer to a new tube, and precipitate with triple volume of prechilled absolute alcohol and invert the tubes for 10 times to precipitate white clumps of nucleic acid.
2. Incubate the tubes at -20 °C for one hour (at this step the tubes can be kept for overnight also in -20 °C freeze).

DNA purification

1. Centrifuge the precipitated sample at 10,000 rpm for 10 min. at 4 °C to get pellet.
2. Wash the pellet by 500 µl 75% ethanol, without shaking the tube, gently swirl and centrifuge at 10,000 rpm at 4 °C for 10 min.
3. Repeat above step with 300 µl at 75% ethanol.
4. Dry pellet at ambient temperature or in dry bath to get rid of smell of ethanol.
5. Add 50 µl of TE buffer and dissolve the pellet by putting tubes in 50 °C water bath for 10 min. and then give a short spin to the tubes.

Quantification of DNA

Quantify the genomic DNA using Nanodrop-spectrophotometer. For quantification, put 1 µl from the stock of genomic DNA on Nanodrop-spectrophotometer and measure the OD (optical density) at 260 nm and 280 nm. The concentration should be measured in ng/µl.

The good quality DNA samples with a ratio of 1.8-1.9 at O.D 260/280 must be used for further down applications. Dilute the stock solution to a final concentration of 50 ng/µl of DNA and use for further down applications.

Observation

____

____

____

____

____

____

References

Doyle, J.J. and Doyle J.L. (1990). Isolation of plant DNA from fresh tissue. *Focus*, 12:13-15.

Student Activities

Activity 1: Laboratory exercise

A. Isolate DNA from given leaf sample and quantify by spectrophotometer.

Activity 2: Answer these questions

Q.1. Explain the importance of CTAB in plant DNA extraction.

Q.2. Explain the role of chilled isopropanol in DNA extraction.

Q.3. Why chilled alcohol is used in DNA extraction?

Q.4. What is the significance of chloroform in DNA extraction?

Q.5. Why deionized water is important in DNA extraction?

Q.6. Write the ratio of OD_{260}/OD_{280} of a good quality genomic DNA?

11

Quality Check of Plant Genomic DNA on Agarose Gel Electrophoresis

Objective: To check quality of genomic DNA on agarose gel electrophoresis.

Background information

Agarose gel electrophoresis is a common method of gel based electrophoresis technique used in biochemistry, molecular biology, genetics and clinical chemistry for separation of Protein and nucleic acids from a mixture of macromolecules. Agarose is a major component of agar that is used as a matrix in gel base electrophoresis. The nucleic acids (DNA and RNA) are separated based on length of fragments while, proteins are separated based on size and/or charge (Kryndushkin, *et al.* 2003). For separation of bio-molecules the electric field is applied that leads to movement of charged particles via agarose gel matrix and bio-molecules are separated based on difference in size in the agarose gel matrix (Sambrook and Russel, 2001).

Gel casting with agarose is easy and has very few charged groups, hence found most suitable for nucleic acid separation in most of the biochemistry, molecular biology and biotechnology laboratories. Separated DNA/RNA may be seen and recorded with stain under UV light. Further DNA fragments can be eluted from agarose gel with an ease for further down applications. The percentage of agarose gel used is in the range of 0.7 – 2. Lower percent of agarose gel is used for long DNA fragments while, higher percent of agarose gel is used for short DNA fragments with suitable running buffer.

In the super coiled bundles the helical agarose molecules are accumulated and form 3-D structure with a lattice of pores and channels, from these pores and channels the DNA can pass(Sambrook and Russel, 2001). The size of pores and channels are decided by the percentage of agarose gel. Three dimensional structure of agarose gel is maintained by weak hydrogen bonds that can be easily broken down by heating and gel may convert in liquid. Melting temperature of agarose gel ranges from 85-95 °C, while gelling temperature ranges from

35-42 °C. Low gelling agarose and low melting agarose can also be prepared by certain chemical modifications that are used in several down applications.

Materials

1. Agarose, 50X TBE buffer (stock)
2. Gel loading dye (6X)
3. Ethidium bromide
4. Apparatus, etc. - A horizontal slab gel tank
   - Power pack
   - UV transilluminator
   - Casting tray and combs

Procedure

1. Quality of DNA should be checked on agarose gel (0.8%) prepared in 1X TBE (Tris 45 mM, Boric acid 45 mM, and EDTA 1mM) containing 5 μl of ethidium bromide (50 μg/ml) in 100 ml buffer.
2. Mix the genomic DNA (5 μl) from stock with 1 μl of 6x BPB gel loading dye and load on the gel.
3. For the separation of DNA run the power pack on 5-6 V/cm^2 of gel size. Visualize the fragments in UV light by transilluminator and scan by gel documentation system.

Observation: (A representative agarose gel (0.8%) containing genomic DNA is shown in fig. 3 for example):

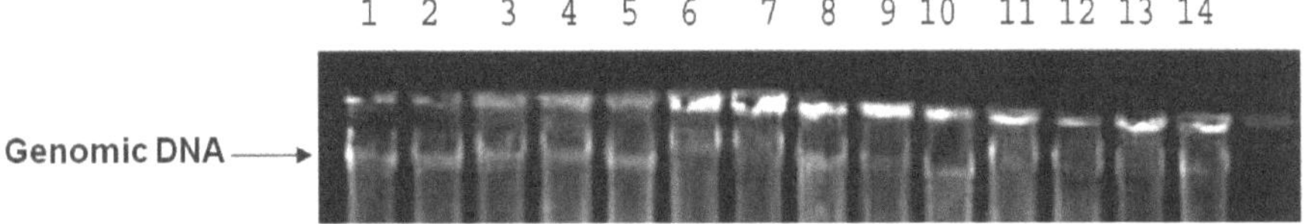

Fig. 3: Genomic DNA of rice leaves (Source: Singh, *et al.* 2017)

References

Kryndushkin, D.S., Alexandrov, I.M., Ter-Avanesyan, M.D. and Kushnirov, V.V. (2003). "Yeast [PSI+] prion aggregates are formed by small Sup35 polymers fragmented by Hsp104". *Journal of Biological Chemistry*. 278(49): 49636–43. doi:10.1074/jbc.M307996200. PMID 14507919.

Sambrook, J. and Russel, D.W. (2001). Molecular Cloning: A Laboratory Manual 3rd Ed. Cold Spring Harbor Laboratory Press. Cold Spring Harbor, NY.

Singh, D., Mehta, R. and Talati, J.G. (2017). Varietal identification of rice using biochemical and molecular marker. Lambert academic publishing, Germany. ISBN: 978-3-330-07905-2, p 87.

Student Activities

Activity 1: Laboratory exercise

A. Check integrity of genomic DNA on Agarose gel.

Activity 2: Answer these questions

Q.1. Why agarose is a suitable matrix for separation of large DNA molecule?

Q.2. Which bonds make 3-D structure of agarose?

Q.3. What should be the melting and gelling temperature of agarose?

Q.4. Explain agarose gel's strength used in molecular biology applications.

Q.5. Write down the type of running buffer used for agarose gel electrophoresis.

Q.6. Write the stain used for visualization of DNA in agarose gel.

12

Randomly Amplified Polymorphic DNA (RAPD) and Electrophoresis

Objective: To amplify DNA using RAPD primers and study on agarose gel.

Background information:

RAPD markers are decamer (10 nucleotide length) DNA fragments from PCR amplification of random segments of genomic DNA with single primer of arbitrary nucleotide sequence and which are able to differentiate between genetically distinct individuals. RAPD is applied to analyze the genetic diversity by random primers. Unlike traditional PCR analysis, RAPD does not require any prior DNA sequence information of the target organism: the identical 10-mer primers will or will not amplify a segment of DNA, depending on positions complementary with primers' sequence. For example, no fragment is produced if primers annealed too far apart or 3' ends of the primers are not facing each other. Therefore, if a mutation has occurred in the template DNA at the site that was earlier complementary with primer, a PCR product will not be produced, resulting in a different pattern of amplified DNA segments on the gel. RAPD is a dominant marker means it cannot distinguish between heterozygous; single copy and homozygous; two copies. Rarely RAPD markers may be Co-dominant where different fragment length of DNA is amplified from the single locus. PCR is a molecular technique used for amplification of DNA, a complete enzyme based chemical reaction, therefore, the concentration and quality of template DNA, concentrations of PCR components, and the PCR cycling conditions may greatly influence the outcome. Thus, the RAPD technique is notoriously laboratory dependent and needs carefully developed laboratory protocols to be reproducible. Mismatches between the primer and the template may result in the total absence of PCR product or decreased amplified product. Thus, the RAPD results can be difficult to interpret.

Polymerase chain reaction (PCR) is a molecular biology tool used to amplify few copies of a segment of DNA across several orders of magnitude, generating thousands to millions of copies of a particular DNA sequence. It is an easy, cheap, and reliable way to repeatedly replicate a focused segment of DNA, a concept which is applicable to numerous fields in modern biology and related sciences (http://learn.genetics.utah.edu/content/lab/pcr).

Developed in 1983 by Kary Mullis, (Bartlett and Stirling, 2003; Mullis, *et al.*, 1983) PCR is now a common technique used in clinical and research laboratories for variety of applications (Saiki, *et al.*, 1985; Saiki, *et al.* 1988). These include DNA cloning for sequencing, construction of DNA-based phylogenies, or functional analysis of genes; diagnosis and monitoring of hereditary diseases; analysis of genetic fingerprints for DNA profiling (for example, in forensic science and parentage testing); and detection of pathogens in nucleic acid tests for the diagnosis of infectious diseases. In 1993, Mullis received Nobel Prize in Chemistry along with Michael Smith for his work on PCR (Mullis, 1993).

The vast majority of PCR methods rely on thermal cycling, which involves exposing the reactants to cycles of repeated heating and cooling, permitting different temperature-dependent reactions—specifically, DNA melting and enzyme-driven DNA replication—to quickly proceed many times in sequence. Primers (short DNA fragments) containing complementary sequences to the target, an enzyme DNA polymerase, that is actually responsible for amplification of DNA repeatedly. After every cycle of PCR reaction the amplified product serve as template for next PCR cycle and thus a chain reaction happens that amplify the original DNA segment exponentially. The basic principal of PCR technique is widely used in for genetic manipulations by extensive modification in chemical reaction. PCR is not a part of recombinant DNA technology because in this technique only DNA fragment is amplified with DNA polymerase enzymes whereas, in recombinant DNA technology the restriction enzymes and ligases are used for cutting and pasting the DNA fragments respectively.

In PCR technique heat stable DNA polymerase is used everywhere, such as *Taq* polymerase, an enzyme originally isolated from the thermophilic bacterium *Thermus aquaticus*. This DNA polymerase enzymatically assembles a new DNA strand from free nucleotides, the building blocks of DNA, by using DNA as template and DNA oligonucleotides (the primers) to initiate DNA synthesis.

There are three major steps in PCR technique. The first step is denaturation step or DNA melting step, in this double helix of DNA is separated in two single strands by breaking down the hydrogen bonds between nucleotides by high temperature (94 °C). Second step is annealing step, where the temperature is lowered that favors the annealing of primers with the single stranded DNA template. The primers anneal where they found complementary sequences on the template. Similarly the single stranded DNA also serve as template for DNA polymerase enzyme; DNA polymerase enzyme selectively starts DNA polymerization where primers are attached. In the third step, the DNA polymerase enzyme starts synthesis of new strand by using denatured or separated single strands of DNA as a template and dNTPs as building block of nucleotide. DNA polymerase starts synthesis of new strand from annealed primers. This step is known as extension. These three steps are repeated for 35-40 times. There will be exponential growth of amplified fragments, e.g. after 40 cycles of PCR there will be 2^{40} copies of specific DNA fragment.

Materials

Components of PCR

(a) dNTPs : 100 mM each, dilute to 10 mM each.

(b) *Taq* DNA polymerase: (3 U/µl).

(c) Primers : RAPD primers (10 p moles/µl)

(d) $MgCl_2$ (25 mM)

Components of agarose gel

(a) Agarose (High EEO type)

(b) 5X Tris borate EDTA (TBE), pH 8.3

0.9 M Tris HCl, 0.9 M boric acid, 20 Mm EDTA.

pH adjusted to 8.3

(c) Gel loading dye (6X)

0.25% Bromophenol blue

0.25% Xylene cyanol

40% Sucrose

Made in distilled water followed by storage at 40 °C

(d) Ethidium bromide (1 mg/ml).

(e) 1 Kb DNA ladder, etc.

Procedure

Amplify the genomic DNA using random RAPD primers. PCR reactions for RAPD should be carried out in a reaction volume of 25 µl. The PCR mix should consist of:

Quantity of PCR components for RAPD

Components	RAPD
PCR buffer (10x) with 15 mM $MgCl_2$	2.5 μl
Primer (10 p moles/μl)	1.0 μl
dNTPs mix (10 mM each)	0.5 μl
Taq DNA polymerase (3 U/μl)	0.3 μl
Template DNA (30 ng/μl)	4.0 μl

Carry out all the PCR reactions in 200 μl thin walled PCR tubes. Tap gently the PCR tubes containing reaction mixture and spin briefly at 10,000 g. The PCR amplification should carry out in thermal cycler subjected to following PCR protocol:

PCR conditions for thermal cycler

Steps	RAPD	
	Temperature (°C)	Duration
Initial denaturation (step 1)	94	5 min
Denaturation (step 2)	94	45 sec.
Annealing (step 3)	40	45 sec.
Extension (step 4)	72	45 sec.
	40 times from step 2 to step 4	
Final extension (step 5)	72	10 min.
Hold (step 6)	4	(Till removal from machine)

Run all the PCR products on agarose gel (1.5 %) containing ethidium bromide (1 mg/ml).

Mix 10 μl of PCR product with 2 μl of 6X tracking dye and load inside the wells on gel.

Run the gel at 80 V (constant) to separate the amplified bands. Also run the standard DNA marker along with the samples.

See the separated bands under UV transilluminator and document by gel documentation system and analyze by available software.

Observation: (A representative 1.5% agarose gel containing amplified DNA fragments with RAPD primer is shown in fig. 4 for example):

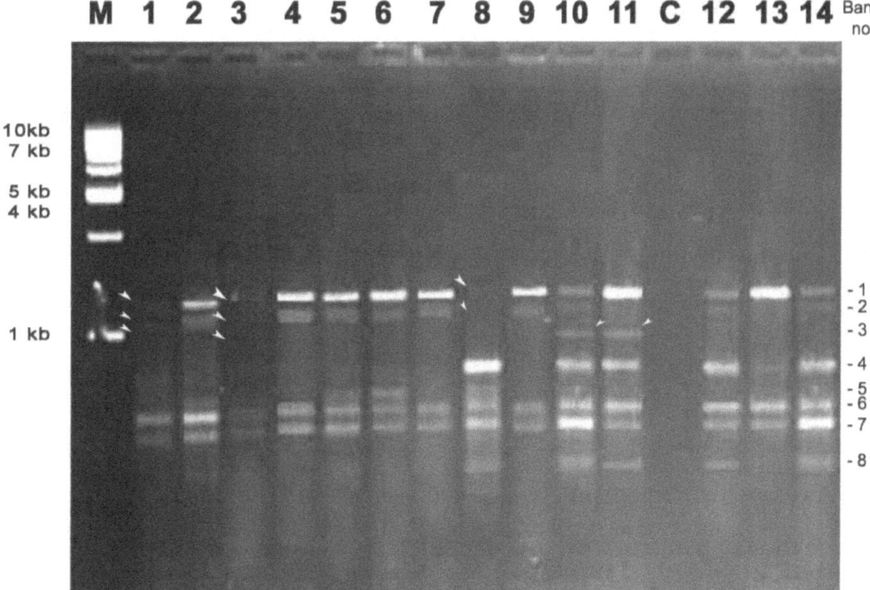

Fig. 4: RAPD profile of rice amplified with RAPD primer OPA 01; M = DNA ladder, 1 to 14 = different samples amplified with RAPD primer and C = Non template control, (Source: Singh, *et al.* 2010)

References

Bartlett, J.M.S. and Stirling, D. (2003). "A Short History of the Polymerase Chain Reaction". PCR Protocols. Methods in Molecular Biology. 226 (2nd ed.). pp. 3–6. doi:10.1385/1-59259-384-4:3. ISBN 1-59259-384-4.

http://learn.genetics.utah.edu/content/lab/pcr "PCR". Genetic Science Learning Center, University of Utah.

Mullis, K.B. (1993) – Nobel Lecture: The Polymerase Chain Reaction.

Mullis, K.B. *et al.* (1983) "Process for amplifying, detecting, and/or-cloning nucleic acid sequences" U.S. Patent 4,683,195

Saiki, R., Gelfand, D., Stoffel, S., Scharf, S., Higuchi, R., Horn, G., Mullis, K. and Erlich, H. (1988). "Primer-directed enzymatic amplification of DNA with a thermostable DNA polymerase". *Science.* 239(4839): 487–491. doi:10.1126/science.2448875. PMID 2448875.

Saiki, R., Scharf, S., Faloona, F., Mullis, K., Horn, G., Erlich, H. and Arnheim, N. (1985). "Enzymatic amplification of beta-globin genomic sequences and restriction site analysis for diagnosis of sickle cell anemia". *Science.* 230(4732): 1350–1354. doi:10.1126/science.2999980. PMID 2999980.

Singh, D., Mehta, R. and Talati, J.G. (2010). RAPD as a molecular tool for identification of rice (*Oryza sativa* L.) cultivars, parents and hybrid. *Indian J Agric Biochem*, 23(1): 27-31.

Student Activities

Activity 1: Laboratory exercise

A. Amplify DNA by RAPD primers and separate by electrophoresis on agarose gel.

Activity 2: Answer these questions

Q.1. Who has invented PCR technique and in which year received Nobel prize?

Q.2. What are different components used for PCR reaction?

Q.3. What are different steps in PCR reaction?

Q.4. What is full form of RAPD and which type of primers are used in this reaction?

Q.5. From which organism *Taq* Polymerase enzyme is obtained?

Q.6. Write the name of instrument applied for PCR reaction.

13

Southern Blotting Experiment

Objective: To transfer DNA on nylon membrane by Southern blotting.

Background information

A Southern blot is a method used in molecular biology to detect a specific DNA sequence in DNA samples. Southern blotting combines transfer of electrophoresis-separated DNA fragments to a filter membrane and subsequent fragment detection by probe hybridization.

The British biologist Edwin Southern (Southern, 1975) discovered this technique the technique is known by his name. Southern blotting technique uses digested genomic DNA with restriction enzymes. Southern blotting technique may be used to identify the gene copies number in a genome. A probe that hybridizes only to a single DNA segment that is uncut with restriction enzyme will produce a single band on a Southern blot, whereas multiple bands will likely be observed when the probe hybridizes to several highly similar sequences (e.g., may be due to sequence duplication).

In the southern blotting technique a short DNA probe is used that matches based on complementary DNA sequence fragment on filter membrane. It proves that the DNA fragment on membrane contain specific sequence for which DNA probe is prepared. When DNA is transferred on the membrane from gel it permits binding of radioactive/Biotin labeled hybridization probe to the restricted DNA. Southern blotting is also important for autoradiography.

Southern blotting is an important molecular biology technique that may be used for homology based cloning, screening a DNA library, screening of cloned DNA fragments and to identify methylated sites in genes.

Materials

1. Plasmid containing gene
2. Transformed plant containing gene (e.g. *PSNV* gene)
3. DNA sequence for probe preparation (e.g. *PSNV* gene)

4. Forward and reverse primers for amplification of known DNA sequence to be used as probe
5. Low melting agarose
6. 1x TAE buffer with 0.5 μg/ml ethidium bromide
7. MinElute PCR purification kit from Qiagen (Cat. No. 28004)
8. Biotin decalabel DNA labelling kit of Fermentas (Cat. No. K0651)
9. Restriction enzymes with compatible buffer (*Nco* I and *Bst* EII)
10. Nylon membrane
11. 6X sucrose loading dye
12. Alkaline transfer buffer
13. Thick blotting paper
14. Filter paper
15. Semi dry blotter (Bangalore Genie, India)
16. Biotin chromogenic detection kit of fermentas (K 0661 and K 0662)

Following solutions are required for southern hybridization

1. 50X TAE Buffer

   For 500 ml : Tris base, 121 g

   : Glacial acetic acid, 28.55 ml

   : 0.5M EDTA, 50 ml (pH 8.0) or add 9.305 g Na_2EDTA

2. 0.5 M EDTA pH 8.0

   For 100 ml stock : 18.61 g Na_2EDTA (mw 372.24)

   Adjust pH 8.0 with NaOH pellets (approx. 2 g required)

3. Alkaline transfer buffer (0.4 N NaOH, 1M NaCl) – Denaturation solution for nylon membrane

   For 500 ml : 8 g NaOH

   : 29.22 g NaCl

   Make volume upto 500 ml in double deionized water

4. Neutralization buffer II (0.5M Tris-HCl – pH 7.2) and (1M NaCl)

   For 100 ml : 6.05 g Tris-base

Adjust pH with conc. HCl to 7.2

Then add 5.844 g NaCl and make up the volume

5. 20X SSC

For 500 ml : 87.65 g NaCl

: 44.10 g Sodium citrate

: Adjust pH 7.0 with NaOH pellets

: Autoclave

6. 50X Denhart's reagents

For 50 ml : 0.5 g Ficol (Type 400)

: 0.5 g Polyvinyl pyrrolidone (pvp), mw 36000

: 0.5 g BSA

: Store at -20 °C

7. Prehybridization solution (0.2 ml/cm^2)

For 20 ml : 6X SSC – (6 ml from 20X SSC)

: 0.5% SDS (1 ml from 10% SDS)

: 5X Denhart's reagent

: Volume make upto 20 ml

8. Hybridization solution (0.2 ml/cm^2), etc.

For 20 ml : 6X SSC – (6 ml from 20X SSC)

0.5% SDS : 1 ml from 10% SDS

: 13 ml double deionized water

: Volume make up upto 20 ml

Procedure

Southern blotting should be done under following steps:

1. Probe preparation

i. In order to prepare probe, amplify the given gene sequence (e.g. *PSNV* coat protein gene) by PCR amplification from plasmid using the gene specific forward and reverse primers.

2. Extraction of DNA from low melting agarose gel

i. Minimum 1000 ng DNA is required for probe preparation.

ii. Prepare one percent agarose gel using low melting point (LMP) agarose in 1x TAE with 0.5 μg/ml ethidium bromide.

iii. Pre-run the gel at 60 V for 10 min.

iv. Loaded 10 μg of PCR amplified product per well with loading dye.

v. Run the gel was at $1V/cm^2$ or less.

Note: Avoid more than 60 V as it can melt the gel.

3. DNA elution from the gel

i. DNA should be eluted from LMP agarose using MinElute PCR purification kit from Qiagen (Cat. No. 28004) and follow the protocol as per manual supplied with the kit.

ii. DNA fragment is excised from the gel with a clean, sharp scalpel.

iii. Weigh gel slice (100 mg) in a clean eppendorf tube and add 300 μl QG buffer supplied with kit.

iv. Incubate for 10 min. at 50 °C or till the gel dissolves completely.

v. Mix the tubes by vortexing every 2-3 min. during incubation.

vi. After the gel is completely dissolved, the color (yellow) of mixture should be equal to the color of buffer QG without dissolved agarose.

vii. Add isopropanol (1 gel volume) to the sample and mix thoroughly.

viii. MinElute spin column is placed a in a given collection tube (2 ml) in a suitable rack.

ix. To bind DNA, the sample should apply to the MinElute column, and centrifuge for 1 min.

x. Discard the flow-through and MinElute column should be placed back in the similar collection tube and 500 μl buffer QG should be added (comes with kit) in the spin column followed by centrifugation for 1 min.

xi. Discard the flow-through and MinElute column should be placed back in the similar collection tube.

xii. In order to wash, 750 μl of Buffer PE must be added to the MinElute column and centrifuge for 1 min.

xiii. Discard the flow-through and now centrifuge the MinElute column for an additional 1 min at 10000 rpm.

xiv. Keep MinElute column in a 1.5 ml microcentrifuge tube.

xv. For DNA elution, add 10 µl buffer EB (10 mM Tris-Cl, pH 8.5) to the center of the membrane and let the column stand for 1 min. and centrifuge for 1 min.

xvi. Estimate eluted DNA (minimum 100 ng/µl and 1.8 quality required) spectrophotometrically with NanoDrop and further store at -20 °C till further application.

4. Labeling of DNA probe

i. Eluted DNA (1000 ng) should be used for labelling by Biotin decalabel DNA labelling kit of Fermentas (Cat. No. K0651) and the procedure should be followed with instructions of supplier.

ii. Add 10 µl DNA template (1000 ng) in a clean centrifuge tube.

iii. In the centrifuge tube also add 10 µl Decanudealide buffer (5X) and make total volume up to 24 µl with distilled water provided with the kit.

iv. Denature the DNA in boiling water bath for 5-10 minute and cool on ice.

v. Then add 5 µl Biotin labelling mix and 1 µl Klenow fragment and adjust final volume to 44 µl with distilled water.

vi. Incubate the above mixture for 1 hour at 37 °C (Incubation can be kept upto 20 hours).

vii. Stop the reaction was by addition of 1 µl 0.5 M EDTA pH 8.0.

viii. Store the labelled DNA probe at -20 °C till further application.

Note: ^{32}P (Radio active isotope of phosphorus) may also used for labeling of DNA probe.

5. Restriction of DNA

i. Restrict the plasmid DNA, Transgenic plant DNA with suitable restriction enzymes (e.g. *Nco* I and *Bst* EII in the practical) as per instructions given in the supplied manual of restriction enzymes.

6. Southern transfer of DNA on nylon membrane from agarose gel

i. Prepare agarose gel (1%) in 1X TAE buffer with 0.5 µg/ml ethidium bromide.

ii. Load approximately 10 μg restricted genomic DNA of positive plant and plasmid DNA or PCR product as a control.

iii. Add 5 μg 6X sucrose loading dye and run the agarose gel (1V/cm^2).

iv. Pre-heat DNA at 60 °C for 2-3 minute before loading.

v. Observe the agarose gel under UV light.

7. Preparation of agarose gel before transfer of DNA

i. Denature the DNA by soaking in denaturing solution (Alkaline transfer buffer) for 15 min at room ambient temperature with gentle agitation.

ii. Repeat above step for 20 minutes with fresh solution.

8. Preparation of membrane for transfer

i. Arrange the filter paper, gel and nylon membrane as shown in fig. 5.

ii. Cut the membrane according to the gel size.

iii. Cut four sheets of thick blotting paper and two sheets of filter paper.

iv. Use forceps to handle the membrane and blotting papers.

v. Float the membrane on the surface of autoclaved distilled H_2O and make it completely wet.

vi. Immerse the membrane for 5 min in the 30 ml alkaline transfer buffer.

vii. Wet the blotting and filter paper in the alkaline transfer buffer.

viii. Assemble the transfer assembly for DNA transfer according to supplier instructions for semi dry blotter.

ix. By rolling a glass pipette remove any air bubbles from between the papers and gel.

x. Run the blotter at 0.8 mA/cm^2 of gel for 1-2 hour [e.g. 6x6 = 36 cm^2, run at 28.8 (30 mA)].

xi. After run, take the membrane and discard the gel (Alternatively stain the gel for 45 min. in 0.5 μg/ml EtBr in H_2O and visualize on a UV for successful transfer of DNA).

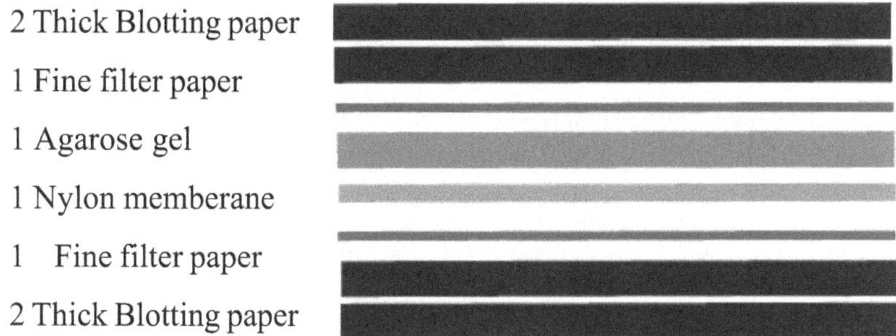

Fig. 5: Arrangement of filter paper, gel and nylon membrane for transfer of restricted DNA to nylon membrane from gel (*Source*: Mehta, 2012).

9. DNA fixation to the membrane

i. For DNA fixation to positively charged nylon membrane, soak it in neutralizing buffer solution II (0.5 M Tris-HCl – pH 7.2 and 1 M NaCl) for 15 min at room temperature.

ii. Now directly use the neutralized wet membrane for hybridization or may be dried for storage.

10. Hybridization of Immobilized DNA and probe

i. Add prehybridization solution with a rate of 0.2 ml/cm^2 of membrane, preheated at 65 °C, and close the bottle tightly (approximately 20 ml) and incubate for 1-2 hours at 65 °C.

ii. Denature the probe for 5 min. at 100 °C and cool it or directly add to the hybridization solution.

iii. Replace the prehybridization solution completely with preheated hybridization solution at 65 °C and add probe (5-10 ng/ml of buffer) in hybridization buffer. (e.g. 200 ng in 20 ml).

iv. Incubate it for overnight at 65 °C with gentle rotation in the hybridization chamber (Thermo Inc., USA).

v. After hybridization wash the membrane with different wash solutions.

11. Post hybridization washes

i. Wash the membrane in first wash solution (2X SSC+ 0.5% SDS) at room temperature for 5 minutes.

ii. Now wash in second wash solution (2X SSC+ 0.1%SDS) at room temperature for 15 minutes.

iii. Now wash in preheated at 65 °C third wash solution (0.1X SSC+ 0.1%SDS).

iv. Finally give a brief wash to the membrane with preheated at 65 °C fourth wash solution (0.1X SSC) at ambient temperature.

v. Now use the membrane for detection of biotinated nucleic acid probes. Care should be taken that the blots may not dry.

12. Detection of probes

i. Detect the probe according to biotin chromogenic detection kit of Fermentas (K 0661 and K 0662). The procedure should be followed as per the manual supplied with the kit.

ii. After post hybridization washes, wash the membrane in 30 ml of blocking/washing buffer (dilute 20 ml of 10X blocking/washing buffer, provided in kit, with 180 ml of double deionized water) for 5 minutes at room temperature on a platform shaker with moderate shaking.

iii. Now block the membrane in the blocking solution (30 ml) (dissolve 0.5 g of blocking reagent, provided in kit, in the blocking/washing buffer; 50 ml) for 30 min at ambient temperature with moderate shaking.

iv. Prepare 20 ml diluted streptavidin-AP conjugate (Add 4 μl aliquot of concentrated streptavidin-AP conjugate with 20 ml blocking solution just prior to use).

v. Incubate membrane in 20 ml diluted streptavidin-AP conjugate for 30 minutes at ambient temperature with moderate shaking.

vi. Further incubate the membrane with 60 ml of blocking/washing buffer for 15 minutes, discard the solution and repeat above step with fresh blocking/washing buffer.

vii. Incubate the membrane with 20 ml of detection buffer (dilute 4 ml of 10X detection buffer, provided in kit, with 36 ml of double deionized water) for 10 minutes and discard the solution.

viii. Now perform the enzyme based reaction with 10 ml of freshly prepared substrate solution (add 0.4 ml aliquot of 50X BCIP/NBT solution to 20 ml detection buffer just prior to use) at ambient temperature in the dark condition.

ix. The blue purple precipitate becomes visible after 15-30 minutes of incubation.

x. For highest sensitivity allow the colour to develop overnight.

xi. To stop the reaction discard the substrate solution and rinse the membrane with double deionized water for few seconds.

xii. Discard the water and air dry the developed membrane to document the results.

Observation: (A representative southern blot image on nylon membrane after hybridization with specific probe is shown in fig. 6 for example):

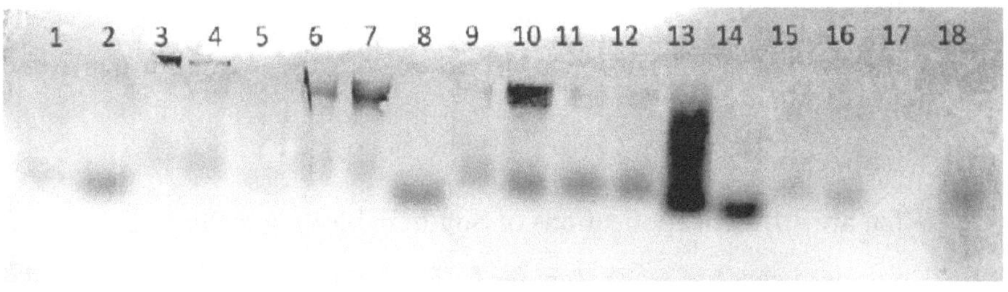

Fig. 6: Southern blot of T_0 plants (Source: Mehta, 2012)

References

Mehta, R. (2012). Development of coat protein mediated transgenic resistance in groundnut to peanut stem necrosis virus. Ph.D. thesis submitted to Faculty of Science, Saurashtra University, Rajkot, Gujarat, India.

Southern, E.M. (1975). "Detection of specific sequences among DNA fragments separated by gel electrophoresis". *Journal of Molecular Biology*. 98 (3): 503–517. doi:10.1016/S0022-2836(75)80083-0. ISSN 0022-2836. PMID 1195397.

Student Activities

Activity 1: Laboratory exercise

A. Perform the experiment of Southern blotting with given DNA sample.

Activity 2: Answer these questions

Q.1. Who has invented Southern blotting technique?

Q.2. What are different steps of Southern blotting technique?

Q.3. Write the name of instrument used for DNA transfer from gel to the nylon membrane.

Q.4. What is probe?

Q.5. What are different applications of Southern blotting technique?

Q.6. How the success of DNA transfer from agarose gel to nylon membrane is verified?

14

Total RNA Isolation from Plants and its Quantification by Spectrophotometer

Objective: To isolate RNA from plants using TRIzol reagent and its quantification by spectrophotometer.

Background information

RNA (Ribonucleic acid) is a polymeric substance present in living cells and many viruses, consisting of a long single-stranded chain of phosphate and ribose units with the nitrogen bases adenine, guanine, cytosine, and uracil, which are bonded to the ribose sugar. RNA is necessary in all the steps of protein synthesis in all living cells and carries the genetic information for many viruses.

The isolation of RNA with high quality is a crucial step required to perform various molecular biology experiment. TRIzol Reagent is a ready-to-use reagent used for RNA isolation from cells and tissues (Chomczynski, 1993).TRIzol works by maintaining RNA integrity during tissue homogenization, while at the same time disrupting and breaking down cells and cell components. Addition of chloroform, after the centrifugation, separates the solution into aqueous and organic phases. Aqueous phase contains RNA.

After transferring the aqueous phase, RNA can be recovered by precipitation with isopropyl alcohol. DNA and proteins can be recovered by sequential separation after the removal of aqueous phase. Ethanol precipitation needs DNA from the interphase, and an additional precipitation with isopropyl alcohol requires proteins from the organic phase. Total RNA extracted by TRIzol Reagent is free from the contamination of protein and DNA. The isolated RNA can be further used in Northern blot analysis, *in vitro* translation, poly (A) selection, RNase protection assay and molecular cloning.

RNA quantification and purity check by spectrophotometer

The traditional and least expensive way to determine RNA quantity and purity is to measure ultra violet absorption of RNA sample using a spectrophotometer.

A_{260} for quantification: RNA has a maximum absorption at 260nm and RNA concentration is determined by the OD reading at 260nm using following conversion: an A_{260} of 1.0 is equivalent to 40 µg/ml of RNA.

A_{280} for protein contamination

In addition to the OD260, measurements should also be taken at 280nm and 230nm. The A_{260}/A_{280} ratio is an indication of protein contamination in the sample. Pure RNA has an A_{260}/A_{280} ratio of 2.1, however values between 1.8-2.0 are considered acceptable for many protocols.

A_{230} for other stuff

In addition to protein contaminants, RNA preparations can also contain contaminants such as guanidine salts and phenol (used in RNA isolation protocols). A high peak at A_{230} indicates contamination with either of these. The ideal A_{260}/A_{230} ratio is greater than 1.5.

Materials

1. TRIzol or TRI reagent
2. Isopropanol (2-propanol)
3. Chloroform
4. 75% ethanol in DEPC H_2O
5. RNase free water (filtered or DEPC), etc.

Note: All labwares, glass wares and polywares should be treated with DEPC.

Procedure

Total RNA isolation should be done under following steps:

1. Cell lysis

i. Take fresh 21 days old leaves (0.1 gm) from each genotype and make powder in liquid nitrogen using a pestle and mortar.

ii. Transfer the resulting powder to a 1.5 ml tight capped eppendorf tube.

iii. Add 1ml TRIzol reagent in it and homogenize by vortexing for 1 min.

iv. Incubate the samples at room temperature for 5 min (5 min at RT for complete dissociation of nucleoprotein complexes).

v. RNA will be stable in TRIzol which deactivates RNases.

2. Phase separation

i. Add chloroform (1/5 volume of TRIzol) in samples and give vigorous hand shake for 15 seconds.

ii. Incubate samples at ambient temperature for 2-3 min.

iii. Centrifuge the tubes at 12000 X g at 2-8 °C for 15 min.

iv. Transfer the top aqueous phase in another eppendorf tube.

3. RNA precipitation

i. Add isopropanol (70% of aqueous phase or 1/2 TRIzol volume) in tubes and incubate at room temperature for 10 minutes.

ii. Centrifuge the tubes at 12000 X g at 4 °C for 10 min.

iii. Remove the supernatant.

4. RNA wash

i. Wash the pellets with70% ethanol (add & vortex briefly) and centrifuge at 7500 X g for 5 min at 4 °C.

ii. Prepare 70% ethanol with RNase-free water.

iii. Air dry the pellet for 10-15 min. (Over drying of the pellet should be avoided).

iv. Dissolve the pellet in DEPC treated water (25-30 µl) and keep at 55-60 °C for 10 min on dry bath.

RNA quantification by spectrophotometer

i. Quantify RNA by Nano Spectrophotometer.

ii. RNA absorbs UV light more effectively and thus make it possible to quantify and detect as low as 2.5 ng/µl concentrations.

iii. In nucleotides the nitrogenous bases have absorption maxima at approximately 260 nm.

iv. The RNA with 1.7-2.1 ratio at O.D. 260/280 should be retained for further use.

Observation

Sr. No.	RNA Sample	OD 260/280

References

Chomczynski, P. (1993). A reagent for the single-step simultaneous isolation of RNA, DNA and proteins from cell and tissue samples. *BioTechniques*, 15:532-537.

Student Activities

Activity 1: Laboratory exercise

A. Isolate RNA from the given sample using TRIzol reagent and quantify by spectrophotometer.

Activity 2: Answer these questions

Q.1. Name the reagent as ready-to-use reagent for RNA isolation.

Q.2. Write the name of chemical used for precipitation of RNA from the aqueous phase.

Q.3. What are different down applications of isolated total RNA?

Q.4. How RNA concentration is determined by the OD reading at 260nm?

Q.5. Write the ratio of A_{260}/A_{280} for consideration of pure RNA.

Q.6. What are different steps of RNA isolation by TRIzol reagent?

15

Quality Check of Total RNA on Agarose Gel

Objective: To check the integrity of isolated RNA on agarose gel.

Background information:

The most common method used to assess the integrity of total RNA is to run RNA sample on agarose gel (denaturing) stained with ethidium bromide (EtBr). While native (non-denaturing) gels can be used, the results can be difficult to interpret. The secondary structure of RNA alters its migration pattern in native gels so that it will not migrate according to its true size. Nondenaturing conditions also result in bands that are not as sharp, and even multiple bands representing different structures of a single RNA species.

Intact total RNA run on a denaturing gel will have sharp, clear 28S and 18S rRNA bands (eukaryotic samples). The 28S rRNA band should be approximately twice as intense as the 18S rRNA band. This 2:1 ratio (28S:18S) is a good indication that the RNA is completely intact. Partially degraded RNA will have a smeared appearance, will lack the sharp rRNA bands, or will not exhibit the 2:1 ratio of high quality RNA. Completely degraded RNA will appear as a smear (having very low mw). Inclusion of RNA size markers on the gel will determine the band size or smears size and will also serve as a good control for ensuring proper gel run. **Note**: Poly(A) selected samples will not contain strong rRNA bands and will appear as a smear from approximately 6 kb to 0.5 kb (resulting from the population of mRNAs, and depending on exposure times and conditions), with the area between 1.5 and 2 kb being the most intense (this smear is sometimes apparent in total RNA samples as well).

Materials

1. 10x formaldehyde agarose gel buffer

   200 mM 3-[N-morpholino]propanesulfonic acid (MOPS) (free acid)

50 mM sodium acetate

10 mM EDTA

pH to 7.0 with NaOH

2. 1x formaldehyde agarose gel running buffer

   100 ml 10x Formaldehyde Agarose gel buffer

   20 ml 37% (12.3 M) formaldehyde

   880 ml RNase-free water

3. 5x RNA loading dye, etc.

   16 μl saturated aqueous bromophenol blue solution

   80 μl 500 mM EDTA, pH 8.0

   720 μl 37% (12.3 M) formaldehyde

   2 ml 100% glycerol

   3084 μl formamide

   4 ml 10 x formaldehyde agarose gel buffer

   RNase-free H_2O to 10 ml

   Stability: About 90 days at 4°C

Procedure

1. Denaturing agarose gel electrophoresis should be done for separation of RNA.
2. Prepare denaturing gel (0.8%) in formaldehyde agarose gel buffer (1.0 X) (Sambrook and Russell, 2001).
3. Dissolve 0.8 gm agarose with 10x formaldehyde agarose gel buffer (10 ml) and add RNase-free water to make 100 ml final volume.
4. Boil the agarose in microwave oven and cool to 65 °C using water bath.
5. Then add 1.8 ml of 37% (12.3 M) formaldehyde and 2 μl of a 10 mg/ml ethidium bromide form stock solution.
6. Mix the gel solution thoroughly and pour onto gel support.
7. Prior to running the gel, equilibrate the gel for 30 min in formaldehyde agarose gel running buffer (1x).

8. Add one volume of 2x RNA loading dye per volume of RNA sample and incubate for 10 min at 70 °C, cool to chilling on ice and load onto the equilibrated formaldehyde agarose gel.
9. Run the gel in formaldehyde agarose gel running buffer (1x) at 5-6V/ cm^2 till the dye reaches three fourth of gel.
10. Visualize RNA fragments using UV transilluminator and document it using gel documentation system.
11. After resolving by electrophoresis the 28S and 18S RNA should exhibit near 2:1 ratio on ethidium bromide staining indicating that no significant degradation of RNA has occurred.

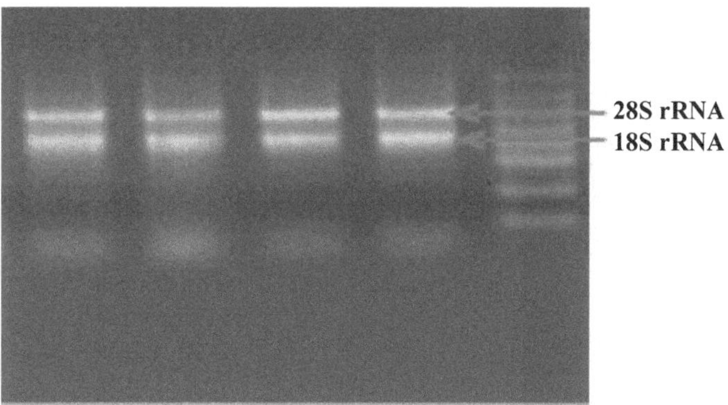

Fig. 7: Total RNA profile showing intact RNA isolated from cotton (*Source*: Prajapat, *et al.* 2016)

Observation: (A representative image of denatured agarose gel showing intact total RNA in 2:1 ratio of 28S rRNA and 18S rRNA as depicted in fig. 7 for example):

References

Prajapat, P., Singh, D., Suthar, K.P., Khushbu, P., Parekh, V.B. and Ahmad, T. (2016). Selection and validation of reference genes in *Gossypium hirsutum* L. under drought stress by quantitative RT-PCR.

Sambrook, J. and Russell, R.W. (2001). Molecular cloning: A laboratory manual, 3rd ed. Cold Spring Harbor Laboratory Press, cold spring harbor, N.Y.

Student Activities

Activity 1: Laboratory Exercise

Check the integrity of isolated RNA on agarose gel.

Activity 2: Answer these questions

Q.1. Which agarose gel is used for checking the integrity of total RNA?

Q.2. Why native gels are not suitable for checking the integrity of total RNA?

Q.3. In case of eukaryotic sample, how many bands should visible in the agarose gel if total RNA is intact?

Q.4. Write the ratio of 28S and 18S for intact total RNA.

Q.5. Write agarose gel percentage used for checking integrity of RNA.

Q.6. What will be visible on agarose gel if RNA is degraded?

16

cDNA Synthesis from Isolated Plant Total RNA

Objective: To synthesize cDNA from isolated plant total RNA.

Background information:

Complementary DNA (cDNA) is double-stranded DNA made from template of single stranded RNA, [e.g., mRNA or microRNA] by reverse transcriptase enzyme. Mostly cDNA is used for cloning of eukaryotic genes in the prokaryotes. For the expression of a specific protein in a cell where the expression of that protein is not found (i.e., heterologous expression), the cDNA will be transferred that codes for the protein to the recipient cells. Retroviruses (e.g. HIV-1, HIV-2, simian immunodeficiency virus, etc.) also produce cDNA naturally and integrate in host's genome and further creates provirus (Croy, 1998).

Major use of cDNA is in gene cloning or as gene probes or in the creation of a cDNA library. When scientists transfer a gene from one cell into another cell for the expression of new genetic material as a protein in the recipient cell, the cDNA will be added to the recipient (rather than the entire gene), because the DNA for an entire gene may have certain DNA that may not code for the protein or that interrupts the coding sequences of the protein (e.g., introns). Partial sequences of cDNAs are often obtained as expressed sequence tags.

Materials

1. cDNA synthesis kit, Fermentas; K1612, etc.

Procedure

1. Synthesize cDNA using cDNA synthesis kit (Fermentas; K1612). In this kit random primers are used initiation of cDNA synthesis.
2. Random primers are effective for first strand synthesis from all RNA species viz., rRNA and mRNA.
3. following method should be used for preparation of master mix for cDNA synthesis:

Preparation of first strand cDNA synthesis

Component	Composition
Template RNA	0.1-1 µg
Oligo(dt)18 primer	1 µl
Water, nuclease free	Variable to make volume upto 11 µl
Total volume	11 µl

4. Gently mix the reaction mixture and briefly centrifuge followed by incubation for 5 min at 65 °C.
5. Cool on ice, centrifuge to spin down and keep the vial.
6. Add following components in the given order:

5X reaction Buffer	04 µl
RibolockTM RNase Inhibitor (20u/µl)	01 µl
10 mM dNTP mix	02 µl
M-MuLV Reverse Transcriptase (20u/µl)	02 µl
Total Volume	20 µl

7. Then gently mix and it briefly centrifuge.
8. Incubate at 37 °C for 60 min.
9. Stop the reaction by heating at 70 °C for 5 min.

Directly use the reverse transcription reaction product for PCR application and preserve at -20 °C.

Observation

____

____

____

____

____

____

Reference

Croy, R. (1998). "Molecular Genetics II - Genetic Engineering Course (Supplementary notes)", Durham University durham.ac.uk.

Student Activities

Activity 1: Laboratory exercise

A. Synthesize cDNA from given RNA sample.

Activity 2: Answer these questions

Q.1. What is template for cDNA synthesis?

Q.2. Which organism produce cDNA naturally?

Q.3. What are different uses of cDNA in molecular biology?

Q.4. Write the type of primer used for first stand cDNA synthesis?

Q.5. Which enzyme converts RNA to cDNA?

Q.6. What is intron?

17

Indirect Competitive ELISA

Objective: To quantify aflatoxin content in given sample by Indirect competitive ELISA.

Background information:

Enzyme linked immuno sorbent assay (ELISA) is a common biochemical technique that uses the principle of solid-phase enzyme immunoassay (EIA) for detection of substances mainly antigens in a liquid or wet sample.

ELISA is a very popular technique that is used for diagnostic purpose in health sector, plant protection and also for quality test in food and other industries. This technique applies the principle of antigen-antibody reaction on a solid surface. The attached antibody is linked to certain enzymes and in the end of reaction by addition of substrate that performs enzyme-substrate reaction and makes product that has some signal mostly color change and based on intensity of color the qualitative as well as quantitative estimation of antigens can be done using ELISA reader coupled with spectrophotometer.

In ELISA technique one antibody should be specific to certain antigen. For reaction of ELISA the ELISA plates are used that is also known as microtiter plate made-up of polystyrene. The sample to be tested for ELISA contain unknown amount of antigen to be detected, this sample containing antigen is immobilized on a solid support like polystyrene microtiter plate in two different ways; first is non-specific where antigens adsorbed to the surface and second is specific where the antigens of sample are captured by specific antibody to the antigen, like used in sandwich ELISA. Once the antigens are immobilized non-specifically or specifically the detection antibody should be added that makes precise antigen-antibody complex. The detection antibody may be either covalently linked to certain enzyme or may be detected by another secondary antibody linked to bioconjugated enzyme. During ELISA steps several proteins or antibodies may non-specifically bound that can be removed by washing with mild detergent solution in between of each ELISA step. For washing of ELISA plates, ELISA plate washer may be used that mechanically wash the ELISA plates. At the end of ELISA the final wash is given and substrate is added that produces visible signal in the form of color change that indicates the presence of antigens in the sample and further quantification may be done using ELISA reader coupled with spectrophotometer.

ELISA can also be used for different types of ligand binding assays. In this technique other than antigens any other ligating reagent can be adsorbed on the polystyrene microtiter plate that can bind specifically with detection reagent and finally enzyme may be signal production and quantification. During washing step only ligand and its binding counterpart are immunosorbed by antigen-antibody interaction to the microtiter plate. The ELISA plates have immunosorbed reaction products that is not washed easily so, these plates may not be reused.

Materials

1. 100g kernel (contaminated and healthy)
2. 70% methanol
3. Mixer grinder
4. Whatman No.41 filter paper
5. PBS- Tween
6. AFB_1- BSA
7. microtitre plates (ELISA Plates)
8. 0.2% BSA
9. Antiserum
10. Goat anti-rabbit IgG with alkaline phosphatase labeling
11. Substrate [15 mg ñ-nitrophenyl phosphate (pnpp)
12. 10% Diethanolamine pH 9.8 (DEA)
13. ELISA Reader, etc.

Procedure

The procedure for analysis of aflatoxin B_1 by ELISA must be followed as explained by Singh *et al.* (2015):

1. Sample extraction

1) Take 100g kernel and grind in a blender to a fine powder.
2) Take 20g of above and add 100ml of 70% methanol and blend in a mixer.

3) Transfer the content in a conical flask and shake at 300 rpm for 30 minute.

4) Filter with filter paper (Whatman No.41) and dilute the extract in PBS-Tween (1:10).

2. Coating

1) Prepare AFB_1-BSA conjugate in carbonate coating buffer at 100 ng/ml (1µl AFB_1-BSA + 10 ml of carbonate coating buffer and vortex).

2) Dispense 150 µl of diluted AFB_1-BSA to each well of microtitre plates (ELISA Plates).

3) Incubate at either 4 °C for overnight or at 37 °C for 2 hours.

4) Wash the plates thrice with PBS-Tween, allowing 3 minutes for each wash (Coating of AFB_1-BSA conjugate on the wall of ELISA plate has been shown in below figure):

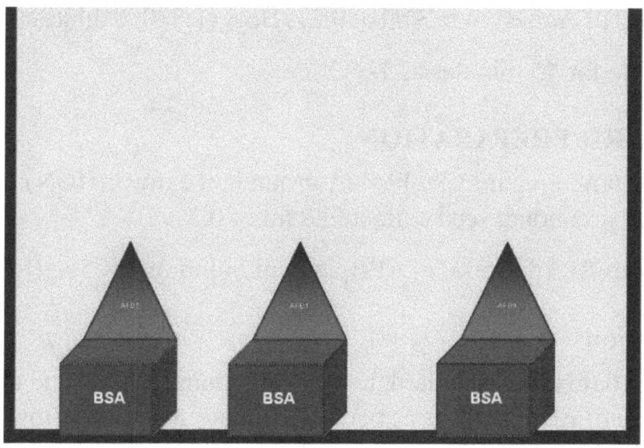

3. Blocking

1) Add 200 µl of 0.2 % BSA prepared in PBS-T (2mg/ml) in each well and incubate at 37 °C for 1 hour.

2) Wash the plates thrice with PBS-Tween, allowing 3 minutes for each wash (Blocking of the empty spaces on the wall of ELISA plate by BSA has been shown in below figure):

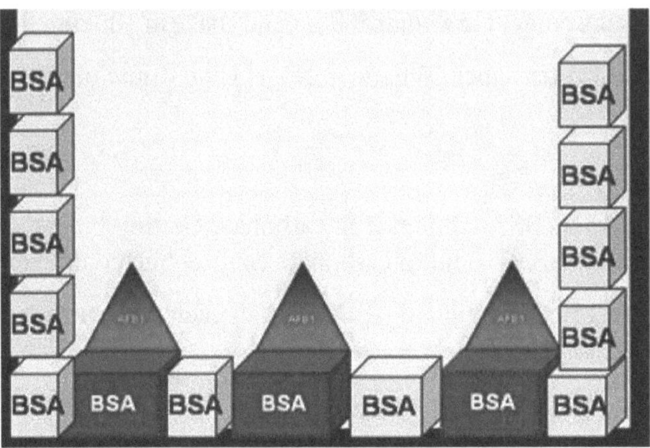

4. Preparation of antiserum dilution and aflatoxin standards

1) Antiserum dilution

i. Add 1µl of antiserum in 3 ml of 0.2% BSA (1:30000 dilution).

ii. Incubate for 45 minutes at 37 °C.

2) STANDARD PREPARATION

i. **Solution A:** Prepare 10% healthy groundnut extract (HGN): Add 1ml of healthy groundnut seed extract in 9 ml of 0.2% BSA.

ii. **Solution B:** Add 25 ml of AFB_1 in 1 ml Solution A (25 ng/100µl AFB_1).

5. Competition

1) Load Aflatoxin B_1 standard, healthy groundnut extract and test samples in wells of ELISA plates as per the template shown in below figure:

2) Add 50 µl of antiserum in every well and keep for incubation at 37 °C for 1 hour (Volume per well will be 150 ml)

3) Wash the plates thrice with PBS-Tween, allowing 3 minutes for each wash (competition between free aflatoxin and bound aflatoxin has been shown in the below figure):

	1	2	3	4	5	6	7	8	9	10	11	12
A					LEA	VE	BLA	NK				
B		100μl Sol B	100μl B + 100μl A	100μl A	100μl A	100μl A	100μl A	100μl A	100μl A	100μl A	100μl A	
C	L E A	100μl Sol B	100μl B + 100μl A	100μl A	100μl A	100μl A	100μl A	100μl A	100μl A	100μl A	100μl A	L E A
D	V E	90μl BSA + 10μl sample	90μl BSA + 10μl sample	90μl BSA + 10μl sample	90μl BSA + 10μl sample	90μl BSA + 10μl sample	90μl BSA + 10μl sample	90μl BSA + 10μl sample	90μl BSA + 10μl sample	90μl BSA + 10μl sample	90μl BSA + 10μl sample	V E
E	B L A	90μl BSA + 10μl sample	90μl BSA + 10μl sample	90μl BSA + 10μl sample	90μl BSA + 10μl sample	90μl BSA + 10μl sample	90μl BSA + 10μl sample	90μl BSA + 10μl sample	90μl BSA + 10μl sample	90μl BSA + 10μl sample	90μl BSA + 10μl sample	B L A
F	N K	90μl BSA + 10μl sample	90μl BSA + 10μl sample	90μl BSA + 10μl sample	90μl BSA + 10μl sample	90μl BSA + 10μl sample	90μl BSA + 10μl sample	90μl BSA + 10μl sample	90μl BSA + 10μl sample	90μl BSA + 10μl sample	90μl BSA + 10μl sample	N K
G		90μl BSA + 10μl sample	90μl BSA + 10μl sample	90μl BSA + 10μl sample	90μl BSA + 10μl sample	90μl BSA + 10μl sample	90μl BSA + 10μl sample	90μl BSA + 10μl sample	90μl BSA + 10μl sample	90μl BSA + 10μl sample	90μl BSA + 10μl sample	
H					LEA	VE	BLA	NK				

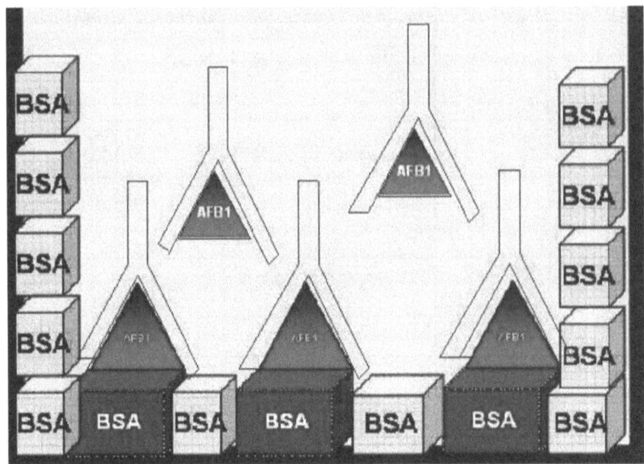

6. Addition of enzyme labeled antibody

1) Prepare 1:1000 dilution of goat anti-rabbit IgG with alkaline phosphatase labeling in PBS-Tween containing 0.2% BSA (10 μl of IgG + 10 ml 0.2% BSA).

2) Add 150 μl in every well and keep for incubation at 37 °C for one hour.

3) Wash the plates thrice with PBS-Tween, allowing 3 minutes for each wash (Reaction between primary antibody and enzyme linked secondary antibody is shown in below figure):

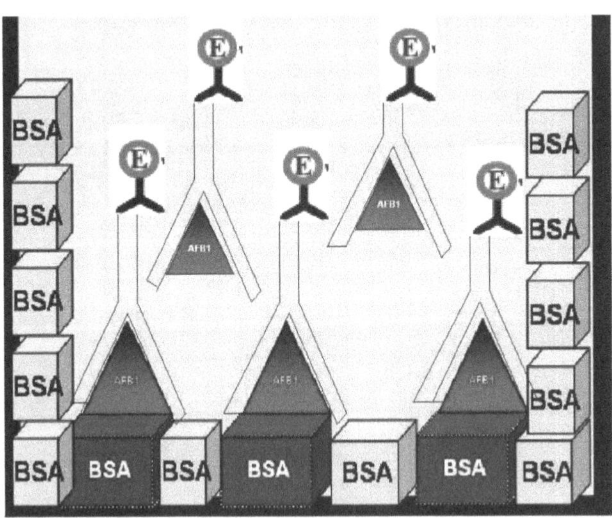

7. Addition of substrate

1) Add 150 μl of substrate [15 mg ρ-nitrophenyl phosphate (pnpp) in 30 ml of 10% Diethanolamine pH 9.8 (DEA)] in each well and incubate for 1

hr at ambient temperature in dark (enzyme substrate reaction will produce yellow color product that is shown in the below figure):

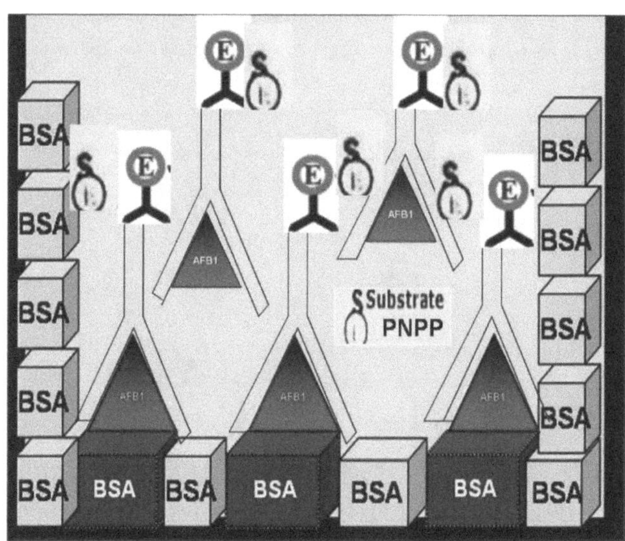

8. Measuring absorbance and calculation of toxin concentration

Measure absorbance at 405 nm in an ELISA Reader

Calculate the aflatoxin B_1 content using following formula:

$$AFB_1\ (\mu g/kg) = \frac{A \times D \times E}{G}$$

A = AFB_1 concentration in diluted sample extract

D = Times dilution with buffer

E = Extraction solvent volume used (ml)

G = sample weight (g)

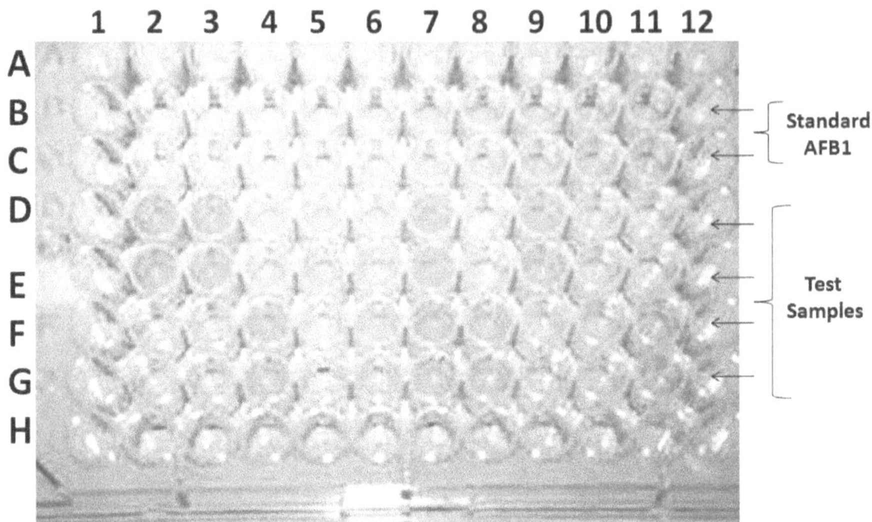

[(B2 and C2 = 100 μl sol. B, i.e. 25 ng AFB1), (B11 and C11 = 100 μl sol. A, i.e. 0 ng AFB_1 or HGN), (B3 & C3 to B10 & C10 shows reducing concentration of AFB_1 as described in the earlier drawn ELISA plate format), D2 to H11 are test samples)]

Note: More yellow color shows less aflatoxin B_1 (AFB_1) and less yellow color shows more aflatoxin B_1.

Observation

(Representative image of ELISA plate after enzyme substrate reaction)

References

Singh, D., Radhakrishnan, T., Kumar, V., Bagwan, N.B., Basu, M.S., Dobaria, J.R., Mishra, G.P. and Chanda, S.V. (2015) Morphological and toxigenic variability in the *Aspergillus flavus* isolates from peanut (*Arachis hypogaea* L.) production system in Gujarat (India). *Journal of Environmental Biology*, 36:441-449.

Student Activities

Activity 1: Laboratory exercise

A. Find out aflatoxin B_1 content in the given sample by ELISA.

Activity 1: Answer these questions

Q.1. What is the full form of ELISA?

Q.2. What is another name of ELISA plate?

Q.3. Why the ELISA plate is washed with mild detergent in the end of each step?

Q.4. In the above experiment write the name of enzyme and substrate that is used for reaction.

Q.5. Are the ELISA plates reusable? If no, then why?

Q.6. What are different steps of indirect competitive ELISA experiment?

18

Objective Questions

1. Which one is the example of abiotic stress?

   a. Drought b. Salinity
   c. Flood d. All of these

2. Which one is the example of osmoticum?

   a. NaCl b. PEG
   c. H_2O d. Prolamine

3. In any given region precipitation below the average, resulting in prolonged shortages in its water supply, whether atmospheric, surface water or ground water is known as _______.

   a. Water use efficiency b. Salinity
   c. Flood d. Drought

4. Which plant family shows maximum drought tolerance?

   a. Fabaceae b. Cucurbitaceae
   c. Poaceae d. Cactaceae

5. Which salts are used to make it saline water?

   a. Sodium chloride & sodium bicarbonate b. magnesium sulfate
   c. potassium nitrate d. All of these

6. Seawater typically has a mass salinity of around ______.

   a. 35 g/kg b. 3.5 g/kg
   c. 0.35 g/kg d. 350 g/kg

7. What is the salinity range of fresh water?

   a. 0-0.5 ppt b. 1-5 ppt
   c. 5-10 ppt d. 10-15 ppt

8. Write the limit of salinity in water used for Agriculture irrigation?

   a. Up to 0.5 ppt b. Up to 1 ppt
   c. Up to 1.5 ppt d. Up to 2 ppt

9. A plant adapted to saline conditions is known as ______.

   a. Bryophyte b. Thermophyte
   c. Halophyte d. None of these

10. Glycinebetaine provides tolerance against abiotic stress by _____.

    a. Osmoregulation b. ROS generation
    c. Chaperon d. All of these

11. Most prevailing salt found in irrigation water is ______.

    a. magnesium sulfate b. Sodium chloride
    c. potassium nitrate d. Sodium bicarbonate

12. In which year Oliver H. Lowry developed the Folin- Ciocalteau reagent for protein assay?

    a. 1930 b. 1940
    c. 1950 d. 1970

13. Which group of amino acid is responsible for color intensity during Folin-Lowry method of protein estimation?

    a. Basic amino acids b. Acidic amino acids
    c. Aromatic amino acids d. Neutral amino acids

14. At which wave length the protein is measured in Folin-Lowry method?

    a. 440 nm b. 550 nm
    c. 660 nm d. 240 nm

15. Who has first time seen the phenomena of electrophoresis (electrokinetic)?

    a. Ferdin and Frederic Reuss (1807) b. Dukhin and Derjaguin (1974)
    c. Laemmli, U.K. (1970) d. None of these

16. What the SDS does in SDS-PAGE technique?

    a. Denatures secondary and non–disulfide–linked tertiary structures
    b. Applies a negative charge to each protein in proportion to its mass
    c. Linearized proteins
    d. All of these

17. Which compound act as cross linker during the process of polymerization in PAG?

    a. Acrylamide
    b. Bisacrylamide
    c. APS
    d. TEMED

18. Percentage of running gel for SDS-PAGE is ____.

    a. 4-5%
    b. 6-7%
    c. 8-9%
    d. 12-15%

19. Percentage of stacking gel for SDS-PAGE_____.

    a. 4-5%
    b. 6-7%
    c. 8-9%
    d. 12-15%

20. ___________ is used for SDS-PAGE.

    a. Horizontal electrophoresis unit
    b. Vertical electrophoresis unit
    c. 2-D electrophoresis unit
    d. Electro blotting unit

21. SDS-PAGE gels are stained with _________ that gives blue color.

    a. Silver stain
    b. Bromophenol blue stain
    c. Coomassie blue stain
    d. Sybr green stain

22. What will be the effect of environmental stress on ROS activity?

    a. Increase slowly
    b. Increase rapidly
    c. Decrease slowly
    d. Decrease rapidly

23. What will be produced by the reduction of molecular oxygen (O_2)?

    a. $^{\bullet}O_2^-$ (Super oxide)
    b. H_2O (Water molecule)
    c. H_2O_2 (Hydrogen peroxide)
    d. $^{\bullet}OH$ (Hydroxyl radical)

24. How many atoms of copper per molecule is present in PPO enzyme?

    a. 2
    b. 4
    c. 6
    d. 8

25. PPO causes the rapid polymerization of _________ to produce black, brown or red pigments (polyphenols) that cause fruit browning.

a. o-diphenols b. monophenol
c. o-quinones d. tyrosine

26. Which ROS enzyme has highest turnover numbers?

a. PPO b. SOD
c. Catalase d. Peroxidase

27. How many amino acids are found in one polypeptide chain of catalase enzyme?

a. 250 b. 500
c. 1000 d. 2000

28. Catalase contains_______ porphyrin heme (iron) groups that is necessary for the enzyme to react with the hydrogen peroxide.

a. 1 b. 2
c. 3 d. 4

29. Optimum pH for human catalase is _______.

a. 4 b. 7
c. 9 d. None of these

30. Which molecule acts as electron acceptor during peroxidases catalyzed reaction?

a. $2H^+$ b. OH
c. COOH d. H_2O_2

31. In plants POX play important role in ________.

a. Construction of cell wall b. Rigidification of cell wall
c. Lignification of cell wall d. All of these

32. Guaiacol will converted into _______ after reaction with POX.

a. Mono guaiacol b. Guaiacol oxide
c. Tetra guaiacol d. None of these

33. Superoxide is produced as a ______ of oxygen metabolism and, if not regulated, may cause cell damage.

    a. Major product b. By-product

    c. Intermediate product d. All of these

34. What is the function of superoxide?

    a. denature enzymes b. oxidize lipids

    c. fragment DNA d. All of these

35. Fe SODs are found in ________.

    a. Chloroplasts b. Mitochondria

    c. Golgi bodies d. Cytoplasm

36. Mn SODs are found in ________.

    a. Chloroplasts b. Mitochondria

    c. Peroxisomes d. Both b and c

37. Cu-Zn SODs are found in ________.

    a. Chloroplasts b. cytosol

    c. extracellular space d. All of these

38. SOD activity should be determined through measuring its ability to inhibit the photochemical reduction of _________.

    a. H_2O_2 b. OH

    c. NBT d. Catalase

39. Enzymes differing in sequence of amino acids but catalyze same reactions are known as ______.

    a. Zymogens b. Isozymes

    c. Antizymes d. Hetero enzymes

40. Allozymes represent enzymes from different alleles of the ____ gene.

    a. Same b. Different

    c.None of these d. Both of these

41. Isozymes represent enzymes from ______ genes.

    a. Same b. Different

    c. None of these d. Both of these

42. For staining of PAGE gel containing peroxidase isozyme ______ dye is used.

a. BPB
b. Coomassie blue
c. O-Dianisidine
d. Methylene blue

43. CTAB is abbreviated for ______.

a. Cetyl trimethyl ammonium bromide
b. Cetyl tetramethyl ammonium bromide
c. Cetyl triethyle ammonium bromide
d. Cetyl tetraethyl ammonium bromide

44. CTAB is a ________ detergent.

a. Anionic
b. Cationic
c. Bivalent
d. None of these

45. Polyvinylpyrrolidone helps in removal of _______ during plant DNA extraction.

a. Polysaccharides
b. Polyamines
c. Polypeptides
d. Polyphenols

46. Which chemical is considered best to use during the extraction/isolation of DNA from plant tissues?

a. CTAB
b. SDS
c. NaCl
d. PEG

47. During DNA extraction the chloroform is applied for separating ____ and _____ from nucleic acids.

a. Proteins and lipids
b. Proteins and polyphenols
c. Proteins and polysaccharides
d. Proteins and polypeptides

48. ________ ratio of Phenol:Chloroform:Isoamyl alcohol is applied during DNA extraction.

a. 25:23:2
b. 25:24:1
c. 25:20:5
d. 25:22:3

49. What is the ratio of OD_{260}/OD_{280} for a good quality genomic DNA?

a. 1.6
b. 1.7
c. 1.8
d. 2.0

50. Most agarose gels are used between _______ percent.

    a. 0.7-1.0 b. 0.7-1.5

    c. 0.7-2.0 d. 0.7-2.5

51. ________ bonds held together the 3-D structure of agarose.

    a. Hydrogen b. Co-valent

    c. Weak interactions d. None of these

52. Melting temperature of agarose gel is ________.

    a. 55-65 °C b. 65-75 °C

    c. 75-85 °C d. 85-95 °C

53. PCR is abbreviated for______.

    a. Polymerase cyclic reaction b. Polymerase chemical reaction

    c. Polymerase chain reaction d. Polymerase cationic reaction

54. PCR technique was invented by______.

    a. Kary Mullis b. Bartlett and Stirling

    c. Beadle and Tatum d. Gregor John Mendel

55. ______ received Nobel Prize in Chemistry along with ______ for his work on PCR.

    a. Bartlett, Stirling b. Kary Mullis, T.H. Morgan

    c. Kary Mullis, H.G. Khorana d. Kary Mullis, Michael Smith

56. Which type of amplification occurs in PCR reaction?

    a. Exponential b. Differential

    c. Integral d. Average

57. PCR technique uses______.

    a. Cutting of DNA b. Pasting of DNA

    c. Amplification of DNA d. All of these

58. What is the function of *Taq* polymerase enzyme?

    a. DNA polymerization at room temperature b. DNA polymerization at lower temperature

    c. DNA polymerization at higher temperature d. None of these

59. Separation of double stranded DNA in single stranded DNA during PCR reaction is called______.

    a. Denaturation
    b. Annealing
    c. Extension
    d. Hold

60. Attachment of primer to the template DNA during PCR reaction is called______.

    a. Denaturation
    b. Annealing
    c. Extension
    d. Hold

61. Formation of new strands of DNA using template of single stranded DNA and dNTPs as building blocks during PCR reaction is called______.

    a. Denaturation
    b. Annealing
    c. Extension
    d. Hold

62. What should be the temperature for denaturation in PCR reaction?

    a. 50 °C
    b. 72 °C
    c. 94 °C
    d. 100 °C

63. What should be the temperature for annealing in PCR reaction?

    a. 40 °C
    b. 50 °C
    c. 60 °C
    d. Depending on primer length

64. What should be the temperature for extension in PCR reaction?

    a. 60 °C
    b. 72 °C
    c. 84 °C
    d. 94 °C

65. The machine used for PCR reaction is technically known as______.

    a. PCR machine
    b. Thermal cycler
    c. Both of these
    d. None of these

66. What is full form of RAPD?

    a. Randomly amplified polymorphic DNA
    b. Rapidly amplified polymorphic DNA
    c. Restricted amplified polymorphic DNA
    d. Rarely amplified polymorphic DNA

67. Generally length of RAPD primers are______.

a. 5 nt

b. 10 nt

c. 15 nt

d. 20 nt

68. RAPD product is resolved on _______ percent agarose gel.

a. 0.8

b. 1.0

c. 1.2

d. 1.5

69. *Taq* polymerase enzyme was originally isolated from ______.

a. *Thermus antranikianii*

b. *Thermus igniterrae*

c. *Thermus aquaticus*

d. *Thermus thermophilus*

70. How much concentration of EtBr is used for DNA staining in agarose gel?

a. 0.1 mg/ml

b. 1 mg/ml

c. 10 mg/ml

d. 100 mg/ml

71. Why DNA ladder should be run in the gel along with DNA sample?

a. To calculate the size of unknown DNA fragment

b. To track the run of DNA fragment

c. To stop the leakage of DNA fragments in buffer

d. All of these

72. RAPD is a _________ marker.

a. Dominant

b. Co-dominant

c. Both of these

d. None of these

73. RFLP is a _________ marker.

a. Dominant

b. Co-dominant

c. Both of these

d. None of these

74. Southern blotting technique was developed by British biologist Edwin Southern in the year ____________.

a. 1970

b. 1971

c. 1973

d. 1975

75. Southern blotting technique can identify _____.

a. DNA

b. RNA

c. Protein

d. DNA and RNA

76. Northern blotting is applied for identification of_____ .

a. DNA
b. RNA
c. Protein
d. DNA and RNA

77. Who has developed northern blot technique and in which year?

a. Edwin Southern in 1975
b. W. Neal Burnette in 1975
c. Harry Towbin in 1977
d. James Alwine, David Kemp, and George Stark in 1977

78. Western blotting technique may be applied for identification of ______.

a. DNA
b. RNA
c. Protein
d. DNA and RNA

79. Who has termed "Western blotting"?

a. Edwin Southern
b. W. Neal Burnette
c. Harry Towbin
d. James Alwine, David Kemp, and George Stark

80. Which technique is used to analyze protein post translational modifications (PTM) such as lipids, phosphomoieties and glycoconjugates?

a. Eastern blotting
b. Western blotting
c. Southern blotting
d. South-Western blotting

81. Carbohydrate epitopes may be detected by.

a. Eastern blotting
b. Western blotting
c. Southern blotting
d. South-Western blotting

82. Which technique is used for detection of protein:protein interaction?

a. Western blotting
b. Far-Western blotting
c. Eastern blotting
d. Far-Eastern blotting

83. Which technique is popular for identification of lipids?

a. Western blotting
b. Far-Western blotting
c. Eastern blotting
d. Far-Eastern blotting

84. Which technique of blotting uses lectin staining?

a. Western blotting
b. Far-Western blotting
c. Eastern blotting
d. Far-Eastern blotting

85. Who has developed the technique Southwestern blotting?

a. Harry Towbin in 1977
b. James Alwine, David Kemp, and George Stark in 1977
c. B. Bowen, J. Steinberg and colleagues in 1980
d. W. Neal Burnette in 1975

86. Southwestern blotting technique detects _____.

a. DNA
b. RNA
c. Protein
d. DNA binding proteins

87. __________ technique may be applied for detection of methylated sites in particular genes?

a. Eastern blotting
b. Western blotting
c. Southern blotting
d. South-Western blotting

88. Ready to use reagent for RNA isolation is _______ .

a. DNAzol Reagent
b. TRIzol Reagent
c. GD Reagent
d. All of these

89. RNA has a maximum absorption at _______ .

a. 260nm
b. 280nm
c. 230nm
d. 240nm

90. Pure RNA has an A260/A280 ratio of _______.

a. 1.6
b. 1.8
c. 2.0
d. 2.1

91. Which type of contaminants may come in isolated RNA?

a. Protein
b. Guanidine
c. Phenol
d. All of these

92. What indicates high peak at A230?

a. Contamination of protein
b. Contamination of guanidine
c. Contamination of phenol
d. Contamination of all of these

93. For a good quality RNA the ideal A260/A230 ratio is _______ 1.5.

a. Lower than
b. Greater than
c. Equal to
d. None of these

94. __________ agarose gel is used for checking integrity of total RNA.

a. Non-denaturing
b. Denaturing
c. Native
d. All of these

95. How many bands will appear on the gel if total RNA is intact?

a. 1
b. 2
c. 3
d. Many

96. Which fragments are visible on the gel after run of intact total RNA?

a. mRNA and rRNA
b. 28S rRNA and 28S rRNA
c. 18S rRNA and 18S rRNA
d. 28S rRNA and 18S rRNA

97. What should be the band intensity ratio of 28S rRNA and 18S rRNA in case of intact total RNA when run on agarose gel?

a. 1:1
b. 1:2
c. 2:1
d. 2:2

98. Conversion of RNA to cDNA is catalyzed by________ .

a. Reverse transcriptase
b. Taq polymerase
c. Restriction enzyme
d. Ligase

99. ELISA is abbreviated for______.

a. Enzyme Limited Immuno Sorbent Assay
b. Enzyme Linked Immuno Sorbent Assay
c. Enzyme Labeled Immuno Sorbent Assay
d. Enzyme Ligated Immuno Sorbent Assay

100. Microtitre ELISA plates are made-up of _______.

a. Polystyrene
b. Polypropylene
c. Polycarbonate
d. Polyvinyl

Zeitfracht Medien GmbH
Ferdinand-Jühlke-Straße 7
99095 Erfurt, Deutschland
produktsicherheit@kolibri360.de